Chas Raymond & James Weir Sewing Machines

By

Alex Askaroff

The rights of Alex Askaroff as author
of this work have been asserted by him
in accordance with the Copyright,
Designs and Patents Act 1993.
©

Sewing Machine Pioneer Series

To see other publications by
Alex Askaroff, visit Amazon.

This is no masterpiece. It's more a self-published labour of love from someone who has spent a lifetime in the sewing trade and a million hours gathering facts for you. Many were sent in by fellow enthusiasts from around the world. I thank you all. If I've missed anyone do let me know.

Why write these books on the Sewing Machine Pioneers? Well, who else bothers? I often feel that if I don't get it all down it will be lost forever. From my vast profits of around tuppence a book I'll invest in a bus ticket to Brighton and scoff candy floss on the pier while chasing seagulls. Please forgive my spelling, United Kingdom English, and enjoy it in the same spirit that it was written.

I have been told that this image of Charles Raymond, used to hang in the lobby of the Guelph Central School, but in recent years it has gone walkabout! It will turn up again but where?

Charles Raymond 6 January 1826 - 4 January 1904

Introduction

It has been a great pleasure researching Charles Raymond all these years. He seemed to be one of the only men who outwitted Isaac Singer and Elias Howe as well as the other big boys engaged in the great sewing machine wars of the 1850's.

In my book on Frister & Rossmann, part of my Sewing Machine Pioneer Series, I go into great detail on how I build my books. It's quite a journey.

Slowly, slowly I have pieced all the information together to bring you a small insight into one of the most successful sewing machine pioneers of the 19th Century. He was knocked back many times but the character of the man was made of stern stuff and, like a prize fighter, each time he was knocked down, he came back stronger.

At his prime he was the largest employer in his area and one of the most respected men of his generation. His efforts help build a city that to this day owes much to this forgotten giant.

I am going to present to you a brief history of the amazing sewing machine pioneer, Charles Raymond. I sincerely hope that you will enjoy reading it as much as I have done researching and writing it. I started in the 1980's and finished it late 2020. No doubt there will be errors. With many facts and figures coming to light almost weekly on the Internet I cross my fingers and hope that I got most of it right.

I'm also aware that there has been some fantastic research on the later Raymond –White models, after Charles Raymond retired. I will leave that story for other enthusiasts and stick to our pioneers in the trade.

Now, I must explain before we get going that our story needs to encompass another great 19th Century sewing machine pioneer, who operated on a different continent. He is inextricably linked to Charles Raymond, one James Galloway Weir.

Weir was the main reason that Raymond machines were such a great success in Europe. He imported in bulk, and sold the machines cheaper than other makers of the period. He marketed Raymond's chain stitch machine as his 55 shilling 'dream machine', and it made him rich.

We will be incorporating James Galloway Weir later on in our story. We will cover a few of the machines that he first imported and later manufactured. We shall also touch on a few other makers as well as we travel along our winding road.

We will also go into some depth on the falling out between these two early pioneers and the huge effect it had on the sewing machine industry. Then, when Weir's story is finished, we shall return to Charles Raymond and wind up his journey

Well, time to get started, buckle up as here we go.

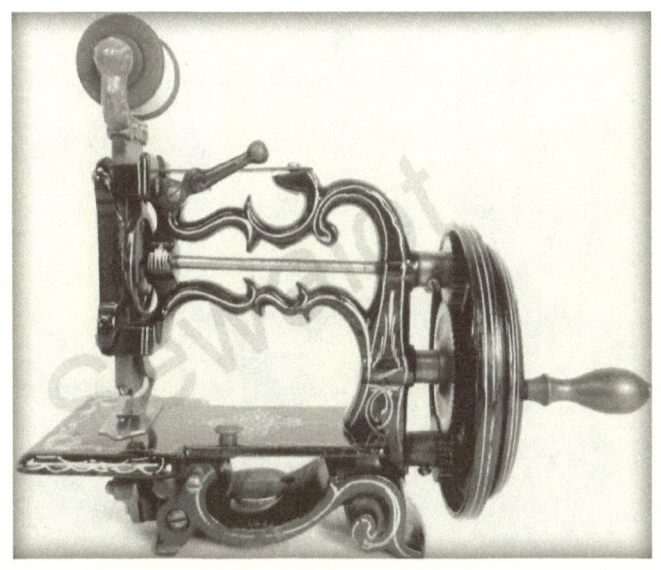

In the 1860's the hand painted Raymond sewing machine was a thing of beauty, and made a good stitch. The first Raymond chain stitch machines that Raymond & Nettleton produced used a simple walking foot and no under-feed.

The needle actually stopped the work moving as the foot slipped over, and then the foot pulled the work forward for the next loop of the chain-stitch, simple but effective engineering. The walking foot was short lived as it needed to be perfectly set to work well. Also the automatic tension on this model often broke.

Because of this, both the auto tension and walking foot were only on the earliest Raymond-Nettleton machines.

The classic hand painted Charles Raymond sewing machine, circa 1861, often called the New England design. There are few prettier sewing machines and all serious collectors hunt these beauties down. As I finish this book in 2020, Harry Berzack has the largest collection of New England machines that I am aware about. I am hard on his tail though!

Chapter One
The Beginning

Charles Raymond was born in Ashburnham, Massachusetts in the winter of 1826. He was the son of a carpenter, Daniel Raymond. His mum was Sarah Greene. With dad a skilled carpenter, joiner and cabinet maker, he would often get Charles to help around his workshop.

This early grounding, between schooling and church, gave Charles the perfect start in tools and machinery.

Charles, called Chas by his friends, left his father's employ at 17 and went to work as an apprentice machinist, at The Massachusetts Cotton Mill Company in Lowell (possibly Boots Mill Museum today).

Once his training was complete, he went onto a daily wage (as a journeyman) where he worked for his living at the Cotton Mill. By the time Chas was 21 he was a well-educated God-fearing man who abstained from drink.

In 1846 Charles opened a machinist's shop in Bristol Connecticut. Here he would carry out work on assorted instruments and mechanical contraptions, making parts when needed. His main trade was the manufacture of clock making machinery.

Of course, with his apprenticeship in the cotton mills, his early experiences working with his father as a carpenter and carriage maker, and his involvements gained from his daily work at his workshop in Bristol, Charles was getting the perfect grounding in the mechanical expertise needed to make a sewing machine.

On Aug 9th 1847, he married Mary Marston who was from the small town of Sharon in Vermont.

Charles was one of the early pioneers in the sewing machine field, and through the late 1840's and 1850's, he saw all the big ideas come to light (and the even bigger egos) from men who were later to become some of the richest in America. Fondly named by the periodicals of the day as 'The Sewing Machine Kings'.

By the 1850's Isaac Singer and many other sewing machine Entrepreneurs were well under way to becoming rich from manufacturing (and copying) all the best ideas that made a good sewing machine. Elias Howe was in the press almost daily, his wealth growing by the second.

It must have occurred to our budding engineer Charles, that there was gold in them hills! Now a skilled mechanical engineer by trade, Charles Raymond was in at the beginning of the American sewing machine industry, and was determined to be part of it.

Charles was used to the intricate mechanisms that produced cloth from his time in the mills, and his clock making experience was also ideal. Like Elias Howe found out, it was a small step from the cotton mills, to the huge leap of mechanically joining fabric together.

From 1846, (the year Elias Howe patented his two great sewing machine patents for the metal shuttle and needle with the eye at the bottom end) until 1851, Charles spent his spare time working on his sewing machine.

In 1851, while still in Bristol, Charles finished his first simple chain stitch machine. An unfortunate coincidence, for it was the same year that Isaac Singer was awarded his first sewing machine patent!

When it came time for Charles to market his single stitch sewing machine, he was stamped on by the rough, tough, Isaac Singer. His idea had to be shelved. The cost of litigation would have destroyed our young entrepreneur.

With no money for patents and no money for court battles, Charles put his sewing machine ideas on the back burner and went back to his clock machinery.

Interestingly these two trades have always been closely connected. The mechanisms for clocks can be seen in many sewing machines. None more so than the Beckwith Sewing Machine. If you ever get a chance do Google it, you will see how huge gears, made on clock making machinery, in Coventry

England, were used to produce a weird looking (but very valuable today) sewing machine.

A few years went by and Charles had not forgotten about his sewing machine. He spent the next few years quietly collecting information for his new venture.

In 1856, Charles had the fortune to meet Willford, Nettleton, a wealthy investor and another maker of clock machinery and parts. Together they prepared themselves for another stab at the sewing machine business in Bristol.

By early 1857 they were ready to get back into making sewing machines. Little did they know that it was going to be another rough ride!

Chapter Two

Patent 17049

Assignors to Henry E Fickett, Glenn's Falls, New York.

To all whom it may concern.
Be it known that we, Willford H Nettleton and Charles Raymond, both of Bristol in the county of Hartford and the State of Connecticut, have invented, made and applied to use certain new and useful improvements in sewing machines.

April 14, 1857

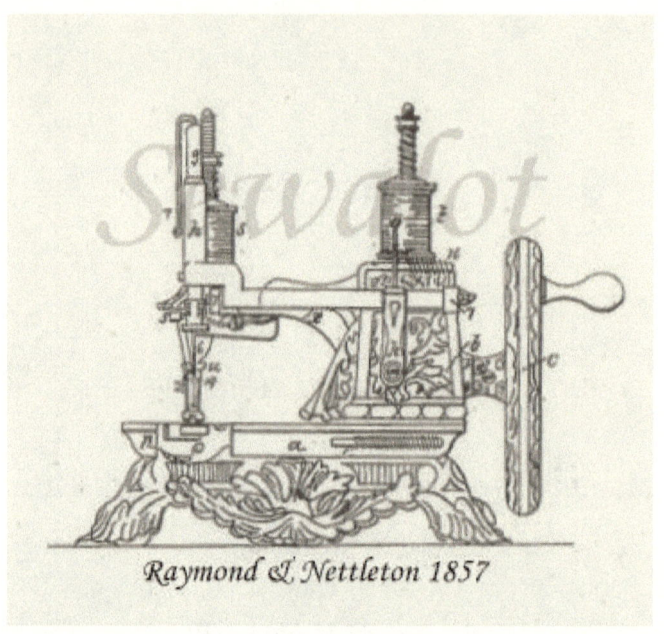
Raymond & Nettleton 1857

April 14, 1857, two thread chain stitch machine Willford H. Nettleton & Charles Raymond. Another discovery trolling through patent applications. The Raymond-Nettleton patent sewing machine of 1857. The ideas went into production but not this beautifully cast and ornate silver-plated design. Only one is known to exist and that must be close to priceless today.

Things moved quickly and by October of 1857 Nettleton & Raymond received their second patent, and in March of 1858, their third patent for a chain stitch machine. Then Nov 30 1858, Patent 22220 for a two thread chain stitch.

By 1858 Charles Raymond formed a legal partnership with Nettleton to produce the basic chain stitch sewing machine that had been patented.

They started manufacturing in an old forge and barn. To make ends meet, while they made sewing machines, they manufactured sewing shears under licence using the patent of J E Hendricks.

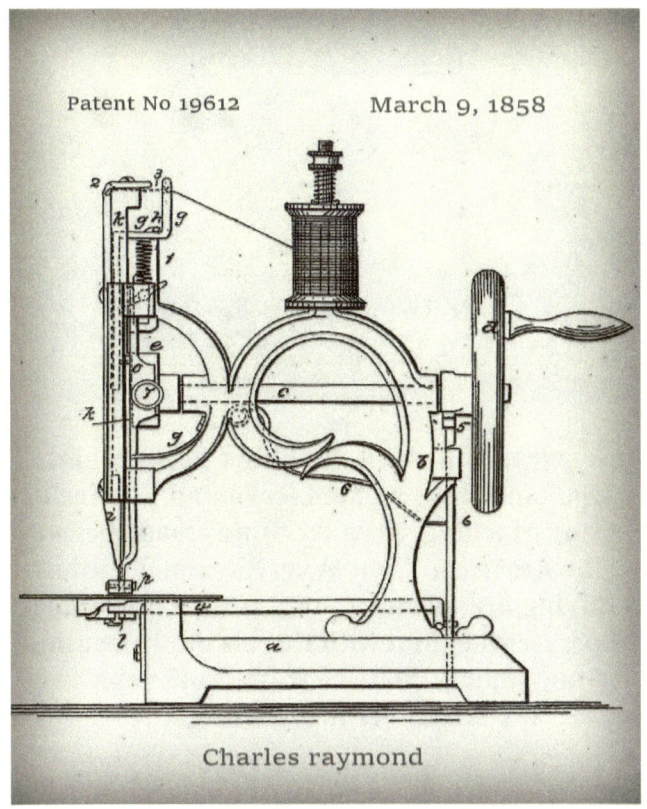

Patent No 19612   March 9, 1858

Charles raymond

Patent No 19612 of 1858. This model has the familiarity of Charles Raymond's later chain stitch machines. Still far more basic than the actual model that he built for sale, but encompassing the necessary details to gain patent protection in America. Still no gears!

It is extraordinary to think that this machine, (similar to his patented Raymond Empire Sewing Machine, circa 1858) was made before the American Civil War. Raymond kept modifying his machines, even while Isaac Singer was persecuting him with threats of prosecution. Many thanks to Mike from Wolfegang's Collectibles for the image.

Around this time, the pair moved from their rented barn in Bristol, Connecticut, to larger premises in Brattleboro, Vermont, where they built a small manufacturing plant. Here they produced a pretty hand painted sewing machine. It did not stitch brilliantly, but it got them a foothold in the expanding sewing machine market.

This is the earliest Raymond-Nettleton in my Sewalot Collection, circa 1858, thinner castings and entirely handmade, possibly by one of the partners or his small team of workmen.

Workers beautifully hand painted the early machines. This meant no two were ever the same. For collectors this is a dream period in sewing machine history.

Image courtesy once again from Mike at Wolfegang's Collectibles. He has probably had some of the finest machines in America pass through his hands over the years.

During the 1850's, hardly any two Raymond Nettleton machines were identical. They were constantly evolving and improving, looking for the perfect machine. It wasn't until 1861, with Raymond's improved looper patent that he finally cracked it. Image courtesy of Mike at Wolfegang's Collectibles.

The machines went on to the market at $14 or even less for bulk buys. The Nettleton-Raymond machine was the smallest and easiest to use, plus it was the cheapest machine on the market. Only a fraction of the cost of a Grover & Baker!

Also remember that 'sewing by machine' was relatively new, so most customers had never seen a lock stitch or a chain stitch.

Imagine you were out shopping in New York in 1858, you see a Grover & Baker machine, it's complicated and takes ages to learn. It is expensive (a year's wages) and heavy. Next you see a Singer which is a big lump of steel and takes ages to thread and sew with.

Then you stumble across Theo Whitfield, agent for the new Raymond machine. It is much cheaper, incredibly easy to use, (needing only one thread), looks beautiful, and seems to join the work just as well. So, you think about it over an early Starbucks and go back to Theo, pay your $14 and take the machine home on your mule. That night by the light of the candle your family watch in amazement as you magically join together fabric by machine.

The future was here! It was this ease of use and price that brought in Nettleton & Raymond big orders.

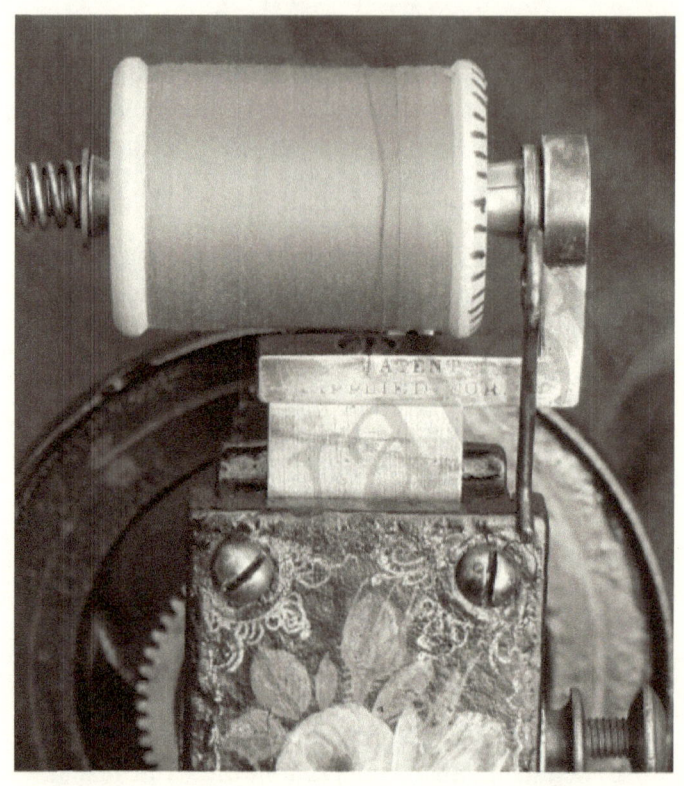

Here you can just make out 'Patent Applied For' showing this model is pre 1861 and possibly one of the earliest Raymond-Nettleton machines to still exist.

Note the unusual reel tensioning bar. By 1861 this was replaced as it wore on the castings very quickly. This is the earliest Raymond in my Sewalot Collection. You can also see the delicate hand painting, more like an oil on canvas.

Only a handful of early Brattleboro Raymond machines survive to this day. This one is in my Sewalot Collection. Notice the complex tensioning bar on the end of the reel of thread. As the needlebar dropped down, the bar pinched the thread tight, to help form the loop. Great idea but it was plagued with problems as it had to be set perfectly to work efficiently.

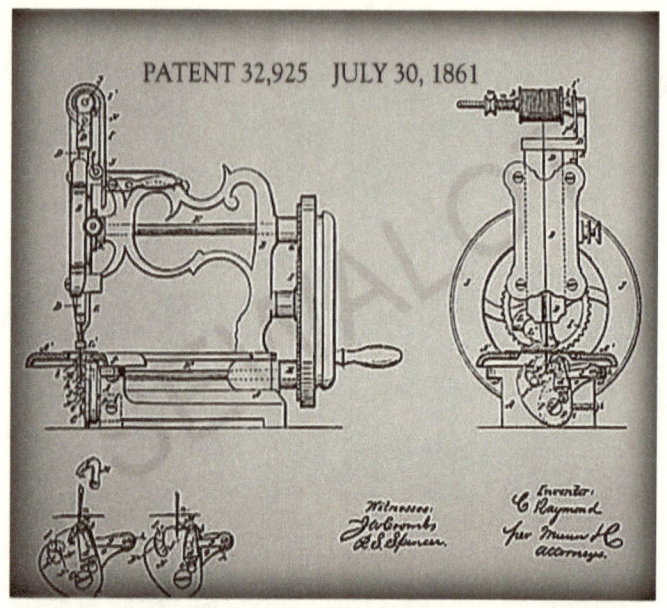

Charles Raymond 1861 patent No 32925, for his chain stitch. The drawings shows the classic Raymond design. However, in 1861 Raymond was still struggling with tension problems.

Note that this patent model does not include the all essential 'thread pinching device' that he later added on the side of the machine. Once he had cracked that, he had the perfect little chain stitch, and the easiest threading sewing machine in the World.

Chapter Three
New England Sewing Machine Makers

I had better quickly explain the term 'New England Sewing Machine'. This term has been applied to the basic shape of the Raymond 'type' chain stitch and others.

There were several manufacturers based around the same area of America all making similar shaped machines. The manufacturers were mainly in the six states, around the New England-Connecticut area of the north-eastern United States. These similar shaped machines became known as 'New England' machines.

It was obvious, that if all the makers were producing a very similar machine (many hand painted so no two were the same) it would be far harder for the big boys to track down the makers and prosecute them for patent infringement!

The manufacturers kept their models clear of marks and production numbers were quite low compared to the larger Singer, Grover & Baker, Wheeler & Wilson, and Howe factories. This allowed a period of a mini boom amongst the New England makers.

Once a machine was sold it was pretty much impossible to tell who had made it and who to prosecute! Remember at this point in American history all sewing machine makers were having to pay the Sewing Machine Cartel or Combination, fees for each model made. The cartel was an illegal monopoly run by the Sewing Machine Kings.

Our problem with all this skulduggery going on, is that it becomes difficult to say who made what, 150 years down the line! Anyway I'll list the makers that I know for certain were producing similar machines to the Raymond Chainstitch.

The Wilson Sewing Machine Manufacturing Company
Cleveland, Ohio

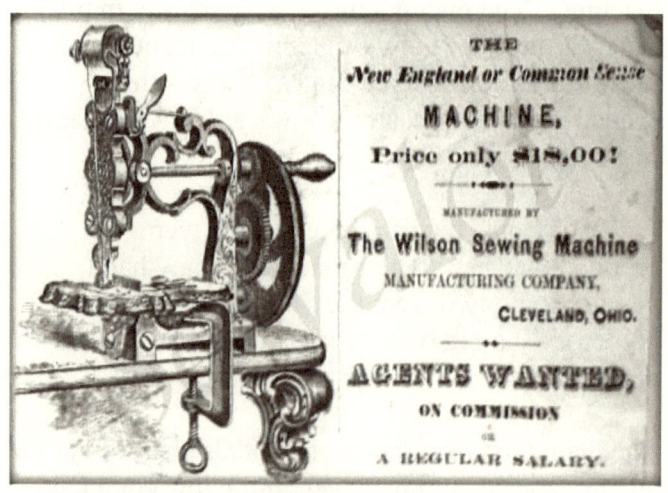

The Wilson New England or Common Sense Sewing Machine

This is one of the several 'New England' clones made by J G Folsom. Notice the lovely hollow-thinner casting similar to the early Raymond models.

The most successful maker, besides Raymond, appears to be Folsom. He managed to set up several times in different locations, outwitting the big boys who were after him for licence fees. He produced a limited number of 'New England' sewing machines until around 1865.

Shaw & Clark
New England Makers, circa 1858-1869
Biddeford Maine.

Shaw & Clark were based in Biddeford Maine and for a few short years, before the legal team of the Sewing Machine Kings caught and prosecuted them, they sold machines infringing on the main patent holders of the time.

However, once they were forced to pay a licence fee for every single machine they made, they used it to their full extent. In their advertising they (falsely) claimed that anyone caught buying a 'non-licensed' New England machine could be prosecuted.

For a few short years Grover & Baker, Elias Howe, Isaac Singer and Wheeler & Wilson ruled supreme as The Sewing Machine Kings. Shaw & Clark were short lived and their machines are extremely rare today, fetching huge sums at auction.

Shaw & Clark sewing machine patent dates

This fascinating badge was put onto all Shaw & Clark machines (after they had paid a licence fee). It shows all the main patent holders from the famous 1846 patent of Elias Howe to the later

1851 Singer, and Grover & Baker patents. Note Elias Howe's name has been included although he had died in 1867. His 1846 patent had to be there for legal reasons. Image kindly sent to me by Rijnko Fekkes.

Those 10 all-important patents from 1846 to 1864, controlled by the Sewing Machine Kings, held back mass-production of sewing machines for over 14 years. Their strangle hold was eventually broken by legal intervention from the government.

This is a rare Shaw & Clark New England model circa 1868, notice the classic wavy edge castings. Also note the pillar design that was used on their later more common 'fire hydrant' models.

Thomas H White & Samuel Barker, Brattleboro (they made the Brattleborough sewing machine).

In 1862, White joined William Grout in Massachusetts to manufacture machines.

Grout & White, Orange, Massachusetts, (Grout later made machines on his own in Winchendon). I know it's a crazy journey. I have images of everyone running everywhere and making little machines being chased by Elias Howe and partners!

J G Folsom, Winchendon, Massachusetts. Bristol, Connecticut and Brattleboro.

Grant Brothers or Grant & Company, 3rd Street Philadelphia.

The Empire Sewing Machine Company. From 1863, they produced a similar New England to Raymond just slightly larger.

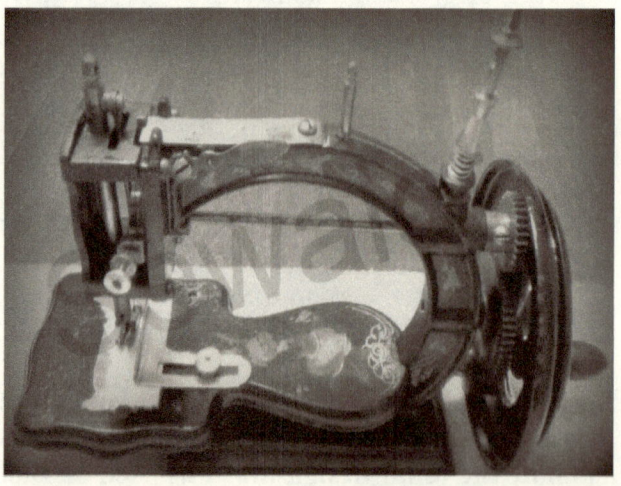

The Grant Brothers New England family had a standard Wheeler & Wilson four-motion-feed with a Raymond looper.

Now back to our old friend Charles Raymond. Charles Raymond possibly had southern sympathies which would explain the seven stars that he put on much of his early advertising.

This may have also been a little dig at the 'Yankee Americans' who, as you will find out, forced him out of his own country into Canada. It wasn't all bad news, the lad later helped to build a great city and made his fortune there.

Chapter Four
Our budding entrepreneurs' journey

Big things were going on in the sewing world in the 1850's. Financiers, money men, and investors saw huge potential in this new-fangled machine that joined cloth quickly.

Many of these investors were cold-blooded big entrepreneurs. They knew that all opposition in business obviously needed to be crushed at birth.

However, by 1858 business was booming for Nettleton & Raymond at their Brattleboro, Vermont foundry and factory. Before long Charles and his pal employed two-dozen workmen and three young lads to fetch and carry and an office girl for the paperwork. She may appear later in our story!

Chas had a great little machine, cheap and easy to use with the simplest threading of any sewing machine yet invented. The adverts were bringing in a good response and yet more expansion was needed.

All this time the machines were improving. If you study the early models you can see the many minor differences as they perfected their little superstar.

The earliest Raymond machines had a thin casting incorporating the needleplate as part of the one-piece bed. By 1860 this was replaced with a removable needleplate. I bought this machine from my fellow collector and friend Maggie Snell. I knew it would be a cracker as she finds some amazing pieces.

*On the Nettleton & Raymond
Use only Geo A Clark, Dixon,
Willimantic, or Cooley's linen thread.
No uneven thread will work well!
Rub hard shaving soap into
Thick cloth for a smooth stitch.*

However, the Indians were circling, and there was trouble at the camp. It was not long before (once again) powerful law suits were brought against Raymond and Nettleton for patent infringement.

Even though some were unfounded, a nice drawn out law suit would close most businesses. The techniques had been used by the Sewing Machine Kings many times, to close down competition that grew too large.

This shows one of Raymond's early patents July 30th 1861. Now with the new removable needle plate but still the early walking foot, no teeth under the plate yet!

The patent for improvements to the looper mechanism of 1861 was stamped on the plates of all the Raymond chainstitch machines from then on. It was his one great patent.

Charles eventually had several patents, all legal and applied for in America, Canada and Britain, but that made little difference to the lawyers whose job was to destroy all opposition by any technically legal means. Even if that meant holding the competition in court for years as their business crumbled.

Unfortunately for Charles, Isaac Singer had patented a similar idea, slightly earlier than Charles, and it was enough to tie Raymond up in expensive litigation.

Note the clever walking foot that simply dragged the work forward. This was only used on the early Raymond machines as it jammed on thicker work.

It looked like their new business was, once again, going to bite the dust almost as soon as it got started.

To avoid the law suits, Raymond and Nettleton had to shut down their business. The big money men probably all lit up cigars poured sippin' whisky and all had a good laugh. But Charles Raymond was far from finished.

Raymond noticed on examination of the patent documents for the court hearings, they were only for America. Canada was exempt, still a virgin territory as far as sewing machines were concerned.

Charles Raymond quietly moved some of his equipment to Montreal. That proved a disaster and almost 30% of his hard-earned investment capital was lost in one badly organised set up.

Charles Raymond then sat down and did his homework properly. This time he found the ideal new town, north of the border where he could produce his beauty.

Chapter Five
Guelph 1861

Guelph in Ontario, Canada, offered Charles Raymond everything he needed and was exempt from patent protection from the big boys down in the lower states.

Guelph was founded on St George's Day April 23, 1827. The growing town was ideal, located beside the Speed and Eramosa Rivers which flow through the town. Originally the town was named to honour Britain's royal family descending from the Guelfs.

The Raymond Manufacturing Company of Guelph, Limited
GUELPH, ONTARIO

John Galt, the Scottish novelist, designed the town to attract settlers. He built the town to resemble a European city with squares and wide streets. Little narrow lanes connected the roads all leading to a grand square, based around the shape of a lady's fan.

Charles Raymond, back in the good old USA, offered his men a new life in Canada if they would follow him and help him start up over the border.

Sixteen key workers moved with Charles Raymond, with only one later returning home. He probably got upset at Charles Raymond's no drinking rule!

In all my research I cannot find any mention of Nettleton at the Guelph factory. It may be around this point in late 1861 that the men parted company.

The Raymond 'Lock-stitch'
Patent 32785
1861

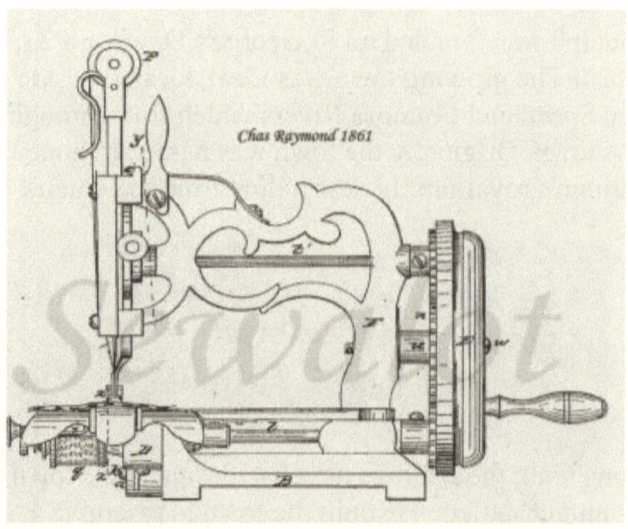

The Chas Raymond sewing machine of 1861. Note the reel of thread under the sewing machine. This was a lock-stitch machine. Patent No 32785. I was so excited when I discovered this patent. Train spotters have nothing on me! Harry Berzack currently knows where one is; it may be unique.

This is a super rare picture of a twin thread Raymond Lock stitch sewing machine. Note the extra reel of thread on the back, going to the hook area. This is far simpler than their original patented two thread chain stitch.

The Raymond Sewing Machine Company

In the autumn of 1861, Charles Raymond opened Guelph's finest sewing machine factory. The Raymond Sewing Machine Company was born. Richard Mott Wanzer (and yes, I do have a book on him as well, in my Sewing Machine Pioneer Series) was already making 1,000 sewing machines a week in his Hamilton factory.

Richard had started the first Canadian Sewing Machine Company back in 1858, and was on a roll. He would later lose almost everything and start again, but his amazing life is a whole different book.

Charles Raymond's later wealth would be heavily invested in Guelph, from the railways to the schools and even street lighting. There is no doubt that he was one of the most influential benefactors to the growing city.

Strangely, although Charles Raymond was a huge beneficiary to the young town, there is currently little mention of him in Guelph's history. Hopefully that will change, as it comes to light all the good deeds he did for the city. I feel a statue coming on!

The early Raymond looper would often catch up the thread and jam. Later improvements cured the problem. Courtesy of Mike at Wolfegang's Collectibles.

Note the improved plate that Raymond used covering part of the looper, developed on the 1861 models. The machines went through constant minor alterations and improvements. No plate, means earlier than 1861.

Charles Raymond's business flourished over the coming years and during the 1860's, competition from America slumped as they fought their Civil War.

His move to Canada had proved to be perfectly timed. All the time Raymond was gathering worldwide patents and slowly stretching out across the Globe.

This was to prove a great success when other companies (trading within their own borders) were hit by the oncoming recession. In Canada his factories grew along with the prosperity of the town itself.

The Raymond Factory on the corner of Suffolk and Yarmouth Street Guelph, Ontario, Canada.

Originally the site of the Arms and Worswick Sewing Machine Co, bought by Charles Raymond in 1875, after a fire destroyed his old wooden factory.

**The Raymond Factory is on the right but to the left of this picture (kindly supplied by Mary Gillett) you can see Charles Raymond's own grand house, complete with white picket fence. Some of the buildings still exist today including Charles Raymond's house.
You can see them through Google Maps, (Norfolk and Yarmouth Streets).**

Chapter Six

Free from most patent litigation, his business boomed and he protected himself with further patents in Canada, America and Europe. He wasn't going to get burnt by Singer again!

Much of his new prosperity was due to his cheap manufacturing. Charles Raymond was able to sell his first chain stitch machine at $14, then reduced it to $10, a fraction of the price of other makes like Singer and Howe machines. Also, his machines were small and light and by now stitched a good seam, although still only as a chain stitch.

Charles Raymond had designed and patented a lockstitch before 1861, complete with a reel of thread under the machine, like the Grover & Baker. But he was doing so well with his little chain stitch that full mass production of the Raymond lockstitch was still years away.

This model from the Berzack Collection has an added skirt to weigh down the machine when sewing. How many of these are original is anyone's guess.

By the 1860's Charles Raymond was on a roll. In 1864, free from persecution from the Sewing Machine Cartel, and with the Civil War raging in America, he had produced over 3,000 chain stitch sewing machines.

By 1870 (around the time this image was taken) they were producing over 10,000 machines a year.

By the 1880's, with over 200 workers, the Raymond Sewing Machine Company had the capacity for over 500 machines a week!

Chapter Seven

In 1869 Mary, his beloved wife, died. His son Arthur, had died when he was very young from diphtheria but his two daughters, Emma A Raymond and Ada F Raymond were grown, and probably proved immensely helpful in getting Charles over this difficult time.

Ada married John B Miner who was a confectioner from Brantford, Ontario. While the eldest daughter, Emma, married John Crowe, who had his own foundry in Guelph. In 1870 Charles built Emma and John a new home on Norfolk Street.

The Crowe Foundry took on all sorts of work for firms that needed castings. I was told that in fact John Crowe had at some point been married to both of Charles Raymond's daughters but I am not sure how that came about? I need to jump into my time machine again and search out these facts! In a few years these statements will be easily cleared up by a quick search on the Internet, but for now it is very peculiar! Someone out there may be able to shed some light on this strange statement.

Love blossomed again for Charles in 1870 when he married for a second time to an old friend, Helen Janet Gill. She was a girl that he had first met in Brattleboro. I wonder if it was his office girl back at the original factory?

To complete his new family, they adopted a son and daughter. Once more Charles Raymond got back to business.

*Light running
Raymond Sewing Machines,
The Best in the World!*

This rare model, the Raymond No1, was another first for the Internet. The picture was kindly sent in by Jane in Nova Scotia in Feb of 2016. It was the first real one that I had seen. It just shows that they are out there! Notice the beaver trade mark. Since online auctions several more have luckily turned up.

*To sew De Laines or other unstarched cloth
Use smooth newspaper underneath.
It will leave you with nice work!
Raymond Sewing Machines.*

Look at this beauty that turned up in France, kindly sent in by Y. Fatin. This is the finest I have seen so far and it hasn't even been cleaned yet!

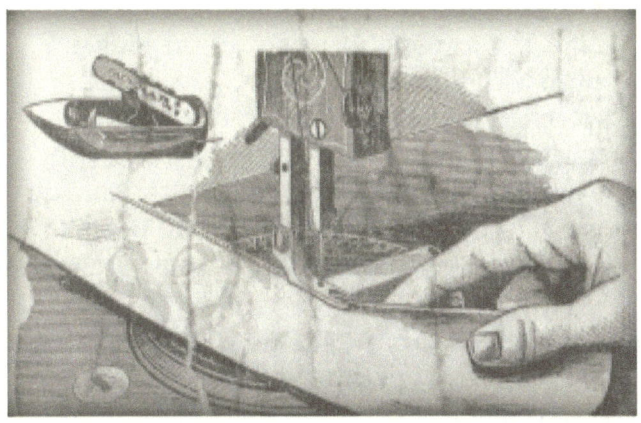

The Raymond No1 lockstitch sewing machine woodcut, with patented shuttle and hemming foot.

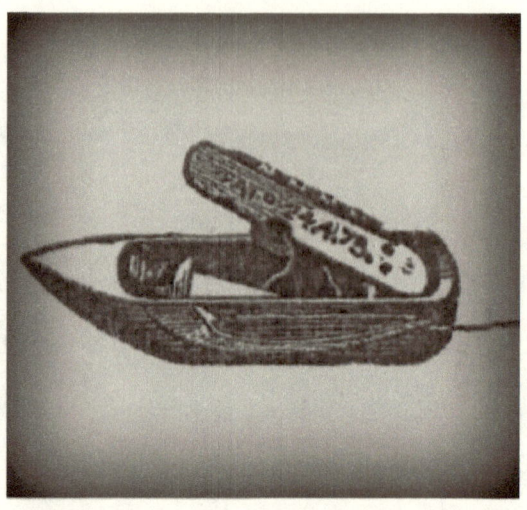

The Raymond Company kept playing with their shuttles, constantly patenting improvements as they went. Their first shuttle machine was advertised in 1869.

They really cracked it with their 1889 patent No 31458. It was possibly a copy of the Singer VS shuttle with minor alterations. The secret to its success was the way the curving shuttle was constantly forced into the bed of the sewing machine as it was swept along by the shuttle carrier. The centrifugal force eliminated any gaps and consequently any missed stitches.

With the help of the latest steam powered machines his factories grew and grew. By 1871 Charles Raymond employed 76 workers including 14 children, also 19 women to hand paint and decorate his machines, with seven in the office.

**There were several versions of this Raymond Household, all highly collectable today.
The Chas Raymond Household sewing machine circa 1870. I sold this machine to a Canadian collector many years ago. I didn't realise till much later, how rare it was! The one I still have is a pale shadow to this beauty.**

Orlando Dunn has been appointed Toronto Main Agent for Raymond Sewing Machines

His business was now turning out over 10,000 machines a year and he was becoming one of the largest sewing machine manufacturers in Canada.

Chas Raymond is currently seeking to employ both skilled and unskilled workers. $6 per week & bonuses (skilled). Apply to the Works Foreman.

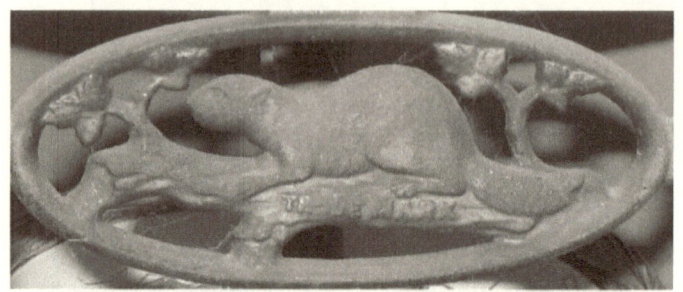

The Raymond Sewing Machine Company Trademark, a wild Beaver. First used in 1872 on his Raymond Household Lockstitch Sewing Machine. The beaver was used by Raymond many years before it became a national symbol for Canada!

In 1872, his skilled workers were earning a dollar a day plus bonuses for reaching production targets. However, production would slow up and bonuses stop during the oncoming Canadian recession, which was a knock on from the American recession.

The beaver was put onto many of Charles Raymond's machines. This one is from his Raymond chainstitch. Legend tells the beaver (and the wealth it brought) was one of the biggest draws to explore Canada. I was amazed

at the trapping of the beaver and its effect on the Canadian economy when I researched my book on the Luton Hat Trade.

Sales to America fell dramatically after a long recession, (starting around 1869). At his factory Charles was now employing over 250 workers and producing nearly a thousand machines a week!

Import restrictions and taxes were imposed by the US, to help home economies. Charles Raymond looked further afield to fill his market. This is where his global patents and agents would pay back big dividends.

Australian Agents
W T Stevens
Ballarat & Geelong
Victoria

Charles Raymond sought out countries all over the world and took on agents and importers, especially in the big European and British markets. The 'old countries' were wealthy and eager for new sewing machines.

Louis Beckh
Quadrate, Markplatz,
Mannheim, Germany.
Importers, Agents & wholesalers of
Raymond sewing machines
1871

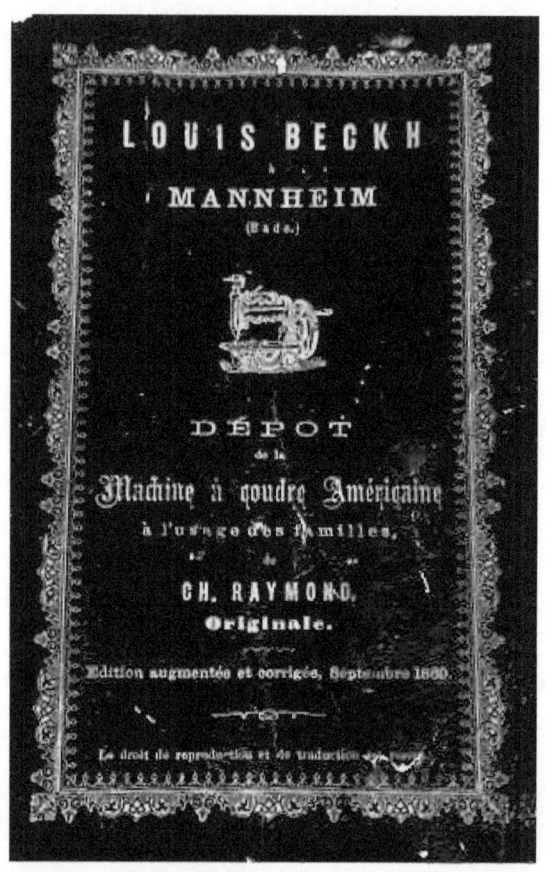

Louis Beckh, Mannheim. Here is a rare cover for the German Raymond imports, kindly sent in by Odile. It shows that Raymond lost no time in expanding his worldwide markets, when his American markets hit trouble.

After two decades, Charles Raymond had finally hit the big time. His Yarmouth Street factories were producing a little cracker of a chainstitch sewing machine which was being widely copied. He had markets and agents in most European countries, and further afield. The order book was full.

John Sheen & Co
Rattray Street, Dunedin
South Island, New Zealand

Main agents for the famous
Raymond Household sewing Machine
Over 5,000 models sold.

Also available,
The Raymond Home Shuttle
Prices from £4.4s.

Some importers even went as far as to have their own extras made for the chain stitch machine. Slippers made (often with the agents' names on) to slip the Raymond machines into (to save clamping to tables). These were sold as optional extras.

I am sure to annoy his American competitors Charles Raymond called his little Canadian beauty 'The American Hand Machine'. Cheeky!

This is a rare sight of an early advertisement for Charles Raymond's first production machine, sold as the Family and Improved Family machine in Canada.

Interestingly in this period of history Sperm Whales were almost hunted to extinction, for their fat used in candles, lamps and sewing machine oil!

Chapter Eight
The Weir-Raymond Connection

**James G Weir 6/07/1839 - 18/05/1911
James was known as Galloway Weir in
Parliament, probably to accentuate his Scottish
roots for his constituency.**

Now, Britain was one of Charles Raymond's biggest markets, outside of North America. In the 1860's, in Britain, Charles Raymond initially dealt with the formidable James Galloway Weir.

For a period, they were inextricably linked, so I must include a little of James Weir's history here.

You'll love it as it's a fascinating tale set around 19th Century London.

I have so enjoyed researching James Galloway Weir. He was full of energy and drive, vim and vigour, also his full share of bluff and bluster too. All-in-all the perfect politician (which he later became). Just a wonderful character to delve into.

For over 20 years in Britain he was one of the most important figures in the sewing machine industry. He loved the tales that surrounded his legend, 'the first British inventor of the sewing machine,' 'a man of untold wealth who circled the elite and royalty alike'. Of course, it wasn't all true, but he did little to expel the inaccuracies. Sewing machines had been around decades before James came onto the scene.

In truth, he did make a fortune from sewing machines. No one is quite sure how much, but then no one is too sure about anything when it comes to James. He was smart enough to get out of sewing machines while the going was good, and by the time he was in his early 40's, he had retired from the bloodbath that was the 19th Century sewing machine business in Britain. He spent a few years wondering what to do and then went into politics.

Hold on to your hats, we are going to go on a roller-coaster ride, trying to track down some of the sewing machine history and legends that surrounds James Galloway Weir.

James was a man on the move and by his early 20's he was already importing machines. At this point in history, Charles Raymond's chain stitch machine was being sold under many names, The Common Sense, The Improved Common Sense, The Family Machine, the Improved Family machine, The Globe, The New England, The Household Fairy, The American Hand Sewing Machine and more!

Often called the Globe sewing machine today. The Weir Globe sewing machine of the 1860's was similar to Weir's other models, some sporting patent 1052, but no other Weir modifications. It was this one machine that made Weir rich.

By 1870, Weir had so many orders for what he called his '55-shilling Dream Machine' that customers would queue outside his London premises in the hope of gaining one. At one point he could list over 100 titled gentry amongst his patrons, including the most powerful person on Planet Earth at the time, Queen Victoria.

The son of a builder (James Ross Weir) James Galloway Weir was born 6 July 1839, one of four children, to James and Margaret Weir.

The entire family moved to London after James finished at the highly acclaimed Dollar Academy, in Clackmannanshire. James was a young man with drive and ambition. By his early 20's he was already importing two sewing machines. From the start of the 1860's, and for the next 20 years he would dominate the London sewing machine scene.

As a travelling salesman, for a haberdashery company, he had travelled the width and breadth of England, constantly meeting customers who needed sewing machines. He knew the potential of a cheap and portable machine. Interestingly he later met his first wife on his travels in Brighton, East Sussex.

His first machine, The Lady, was a German imported chain stitch (not the William Campion Lady). The machine, possibly made by Schroder in Darmstadt, was expensive and problematic.

James was looking for a more reliable and cheaper machine. He knew there was a huge potential for a cheap machine in the expanding Victorian market. At the time expensive and complicated lock stitch machines were dominating sales in Victorian England. He found a beauty in Canada, made by Charles Raymond, which would make him rich!

The later Weir Globe sewing machine of 1873, pretty identical to Weir's other models but sporting patent 944 and 1052. The Globe was a name Weir once used on his unmodified stock. It was offered at a knock down price of 40-45s and did not have the upgrades of his improved machines that sold for 55 shillings.

However, you could return any early machine and have it modified for the sum of one guinea. I doubt if many did as that was half the cost of the machine.

Laws prohibiting how you advertised your wares were scarce and hard to enforce in the 1860's. In 1863 Weir was already set up as an importer or commission agent.

James had spent some time in Canada and had struck up a relationship with Charles Raymond. It was only natural that he saw the potential of a business relationship.

The Raymond machine was to be one of many sewing machines that Weir sold under his own name. To begin with Weir wanted something simple and no sewing machine came any easier to use than the Raymond chain stitch. I mean, it was advertised as the simplest sewing machine in the world!

When Weir started selling the Canadian Raymond, he also advertised the machines in Britain as 'The American Hand Machine', though they were from Canada! We can see from this that James did not mind sailing close to the wind with the facts.

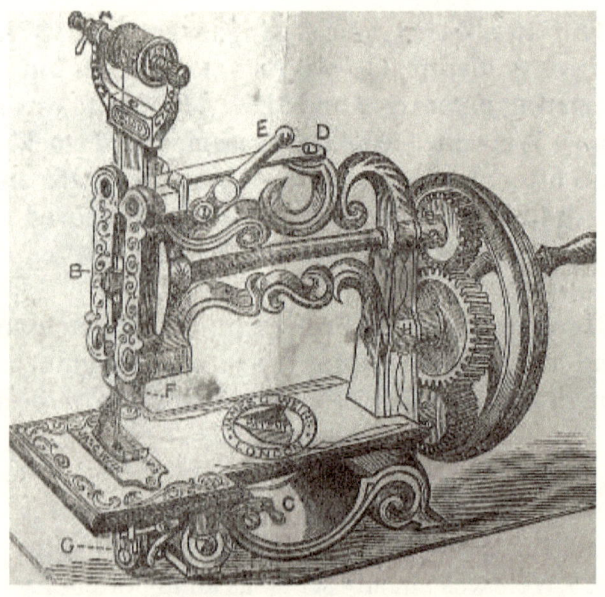

This is a woodcut of the later improved Weir sewing machine. On Nov 29, 1869 Weir also registered the design for the box that the machine went into. It was guaranteed to survive postage to the colonies!

Chapter Nine

And so, by the early 1860's the British Weir sewing machine business was up and running. Earlier than people realise. His later advertising states 1863.

James was living in London at the time. He had premises at Hanaway Street and later a plush retail premises at No 2 Carlisle Street in Soho. Soho was once a centre of the sewing trade.

Soho of course is now far more famous for its shady nightlife, strip clubs and gambling joints rather than long forgotten sewing machine magnates. If you want an exciting night out in London...say no more! In a very short period, with supplies secured, Weir's 55-shilling dream machine became a great success. Within ten years Weir went from sleeping under his workshop bench, too tired to travel home, to become a wealthy man.

They say because of his early struggles in life he was always kind to those with little. His small, light, pretty and simple machine, that produced the most fundamental of all stitches, was making him loads of money. Lucky fella!

The latter half of the Victorian period was one of great invention and discovery and Weir was there to seize the opportunity.

In Britain, it was not until improvements to the Trades Description Act of 1890, that people were banned from stating they made an item that they in fact, just imported. Many importers got away with false descriptions until 1890. One fascinating model came from The Royal Sewing Machine Co, Birmingham.

This amazing machine was won by Daniel Parks back in 2002, it now lives in North Carolina, USA. The machine was simply advertised as a small cast iron sewing machine. Notice the old-style square cut gears which used to chip, wear easily, and were noisier in use.

Sometime between 1878 and 1882 a beautiful copy of the Raymond-Weir came onto the market. The Royal Sewing Machine Company of Small Heath, Birmingham had been originally formed in 1868 by Thomas Shakespear (no 'e' on the end of his name) and George Illstone. Many of their Birmingham sewing machines have the Illstone patent on the shuttle mechanism.

Although Daniel's machine is clearly marked with Shakespear decals, it is a long way off their original Shakespear model, which was a far larger lockstitch machine, as seen below.

The original Shakespear lock stitch Machine

We know the Raymond copy, that they called Shakespear, was manufactured between 1878 and 1882. How? The clue is that it is clearly marked The Royal Sewing Machine Company.

In 1882 the company changed names, becoming the Royal Machine Company (dropping the 'sewing'). Therefore, the machine must have been made between the years that the Royal Sewing Machine Company was formed, in 1878, and its change of name in 1882. It had diversified in manufacture from just sewing machines to several items and carried on until 1888.

Having full manufacturing capabilities in their Herbert Road factory, making a copy of the Raymond machine would have been relatively easy for them. Unfortunately, what we will probably never know is the connection, between Raymond, Weir, and Thomas Shakespear. Did the Birmingham Company make the machines to order? Did they make machines to sell, after seeing how popular the Raymond model was, or did they even make it for James Weir?

Interestingly you can clearly see the straight cut gears and tension assembly that was present on the early Raymond models. Patents, protecting the machine would have run out by the late 1870's.

The Royal Sewing Machine Company tried to market the machine as they clearly state that they displayed the machine at Stationer's Hall, which was a popular exhibition centre in Ave Maria Lane, near Ludgate Hill in London. The Grade I listed building still stands to this day and, amongst other things, is a venue for posh weddings.

The big question is, did The Royal Sewing Machine Company import, like James G Weir, then add their own gold decals, or did they make a near perfect copy of a 'patent free' Canadian Raymond chain stitch to sell, even to Weir down in London?

We may never know. James Galloway Weir was a secretive person when it came to business. He was one of the few that made a fortune from the sewing machine trade. When he retired early, and went into politics it was rumoured that he may have made as much as £145,000 from his businesses. Countless millions in today's terms.

Let's go back to the start of the Weir Raymond trade. For nearly nine years, Raymond and Weir had the perfect business together. Weir's chain stitch sewing machines, made by Raymond, were best sellers BUT by the 1870's Raymond and Weir were bitter business enemies.

After his split with Raymond, James Weir needed regular supplies to feed his ever-expanding market.

Tales say that he had found a French manufacturer (who was already making bases for him) and asked them to produce complete machines. Whether Weir actually imported from France or manufactured these machines himself at his London addresses, is still unclear.

We do know he made large profits which would come from cutting out other suppliers and middle men. More of the French makers later.

Now we can surmise a little here and ask ourselves did the split occur between Raymond and Weir, because Weir was actually setting up and manufacturing the Raymond machine? Probably.

I can imagine that waiting for supplies from Canada must have been tedious, especially when Weir had a booming local market. Making his own simple machine was the best possible scenario. When he actually started making Raymond copies is shrouded in mystery. I bet it evolved rather that started with a bang.

Also buying machines from Canada cost him over a guinea a machine. If he was selling them for around 40 shillings he was not making anywhere near as much money as if he produced the complete machine himself. Details from the period tell us that Weir could manufacture a complete Raymond copy for about 12 shillings! Let's see what we know.

Firstly, we know that he would alter the tension assembly, then the helical gears, then the extra slippers and attachments, then the packing boxes and feed mechanism. Within a few years the only thing he wasn't making was the body of Raymond's machine! We know that he applied for at least four patents and at least two were successful. This tells us that James was no stranger to working on and developing sewing machine mechanisms.

(Shewing the Cabinet Cover removed and Drawer open.

WEIR'S
REGISTERED SEWING MACHINE CABINETS.
(Registered according to Act of Parliament.)

* Polished Mahogany Cabinet, fitted with Lock, Brass Handle, Spring Fastening, and Secret-spring Drawer ; the bottom of the case forming a stand for the Machine, thus avoiding screwing to the table, and keeping it perfectly steady without the addition of extra weight......	15s. 0d.
Polished Stained Pine ditto (to represent ebony)	10s. 6d.
Polished Walnut ditto, very ornamental	18s. 0d.
Polished Mahogany Cabinet, without registered base	10s. 0d.

No extra Charge for Packing Cases for Cabinets.

These Cabinets are invaluable to Ladies travelling, for keeping the Machines clean, and protecting them from children and servants.

ALL ARE OF THE BEST LONDON MAKE AND FINISH.

WHEN A CABINET IS NOT ORDERED, A PLAIN PACKING BOX IS NECESSARY (NOT RETURNABLE)	1s. 0d.

* This Cabinet is specially recommended for Export, and guaranteed to stand any Climate.

Weir put Raymond machines into wonderful boxes. The best was mahogany with a secret drawer, locked to keep those pesky servants out of course! The boxes were brilliantly designed and once clamped securely inside they could be sent all over the world without fear of damage. Note that they were advertised as made in London. Slowly, slowly, Weir was building a machine part by part.

To me it felt like Weir was making a jigsaw puzzle, but not all the pieces. It was obvious that he would eventually copy Raymond's machine completely. Maybe he just imported while he was busy trying to perfect his copy!

James Weir took this split with Raymond as an opportunity to carry on with his sewing machine improvements, namely the gears and the thread holder tension assembly.

K R Gilbert, of the Science museum, kindly supplied this information.

Possible improvements were based on the original Frederick William Parker patent of 1859 (No582). Parker, a Sheffield man, also re-designed and improved a tension device in the same year that helped the stitch no end. Both appeared on the Weir sewing machines.

(The Illustrated London Almanac 1871)
Patronised by Queen Victoria!
The improved & patented Weir sewing machine. One month's free trial!
With improved mesh gears. Still only 55's.
Beware inferior imitations for they are numerous!

Chapter Ten
The Weir Manufacturing Company

By 1870, Weir was publicly advertising as the manufacturer of the 'celebrated' 55-shilling chain stitch, with his London manufacturing base in the Chalk Farm area of NW London, and warehousing in Ferdinand Place.

The Weir Manufacturing Co
Belmont Street,
NW1, London

His prestige offices and showrooms at 2 Carlisle Street, Soho remained his centre for retails sales throughout his sewing machine period.

Weir had modified Raymond machines from Canada using his patented improvements of helical gears, and tension adjusters for his 55-shilling machine. Machines with his improved tension unit and needle slide, were also numbered. For the first time it is possible to start dating early Weir-Raymond machines.

The first Raymond machine he imported and did not modify, he often sold as The Globe machine (and several other names). Unmodified machines sold for the lower sum of around 42 shilling and sixpence.

We also know that in the early years of his business he personally fixed many of the faulty machines in his own workshop. This was the perfect way to learn what was wrong, and how the Raymond machines could be improved.

Automatic Machinery Company Limited

By 1876 Weir went public with his manufacturing, trading as the Automatic Machinery Company Limited. Boasting that his company could produce sewing machines of all varieties, to all trades.

If it wasn't already over, this would have been the final death knell for his business relationship with Charles Raymond in Canada.

Around 1877 he dropped **The Lady** model from Germany, continuing with the cheaper **Globe**, basically an identical machine to the **Raymond New England** type he was previously selling. The decorations were slightly different, but very little else, oh except the price it was now slightly lower at two guineas!

The same year Weir launched the **Zephyr (£4.4s)**, and the **Argus sewing machines**. He already had the **Comet at £4**. He now had a formidable range, but it was still his little improved Raymond 55-shilling dream machine that sold like hot cakes.

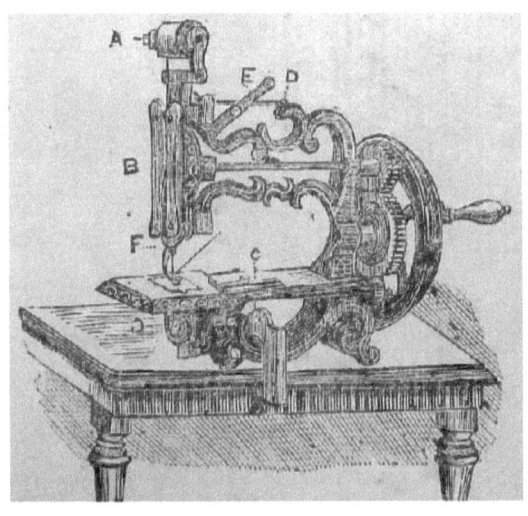

The two guinea Raymond-Weir, and below a woodcut of the different 55 shilling Weir model.

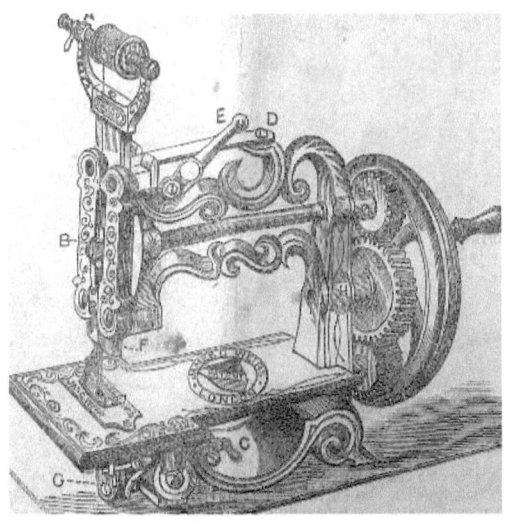

No Weir was genuine without his bed-stamp! Notice the cross-cut helical gears for smoothness, each machine now had a serial number. His improved tension adjustment was a dream too.

No sooner had James Weir gone public with his manufacturing of Raymond copies, he decorated his machines accordingly. They were better than Raymond machines in every way. Weir had listened to his customers and made all the improvements necessary to provide the perfect chain stitch sewing machine. They looked great too. I wonder if Buckingham Palace still has Queen Victoria's in some dusty cupboard.

James Galloway Weir was now publicly advertising Queen Victoria's patronage on all his machines. It's clear that he was still importing some, but also making his own at last. He may have been getting parts from France, Germany, Ireland, and middle England, but his workforce in London was now building machines too.

Weir had reached a pinnacle in his sewing machine career. His machines were now by **Royal Appointment,** after Queen Victoria commanded to see one. "Bring one round young man, and be sharp about it."

He had also supplied H.R.H Princess Mary and a whole list of important establishments, including the Royal Medical College, Guy's Hospital, and my personal favourite, Broadmoor Lunatic Asylum! One can only guess at why they would have needed some, a rush on strait jackets perhaps!

James Weir also listed nearly 100 other 'Distinguished members of the aristocracy' on his adverts. I doubt if they all sewed but, certainly they had staff that did, and if a free machine arrived at the tradesman's entrance, it was a nice perk.

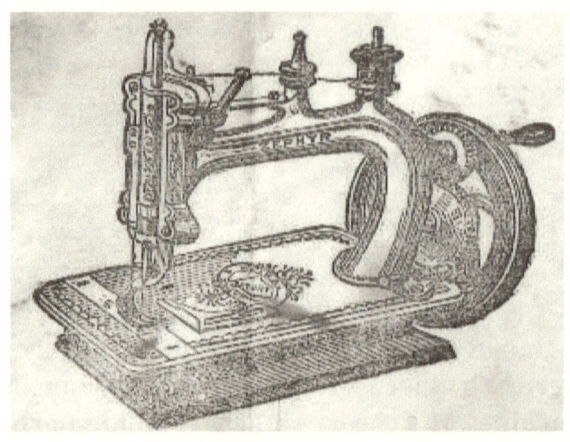

The virtually unknown Weir Zephyr lock stitch sewing machine sold by Weir.

His improvements to the original Raymond machine are more than many people realise. The super-rare Zephyr had many similarities to his chain stitch models. Close examination of one will show how Weir progressed from one machine into the other. This one, with the benefit of a two-thread lock stitch mechanism.

**James Galloway Weir, Zephyr sewing machine.
Circa 1871-85**

*G J Baynes
Agent for Weir Sewing Machines
61 High Street
Gravesend*

Note on this super-rare Weir Zephyr lock-stitch, the similar gearing, presser bar spring, foot lifter arm. They look so similar to the Weir chain stitch that it is easy to see how James' mind was working when he helped designed this machine.

This beauty came onto Ebay in Sept 2007. Another surfaced in 2016. The seller was kind enough to grant me permission to use one of his pictures to show other collectors these super rare delights.

Chapter Eleven
Machines & Improvements

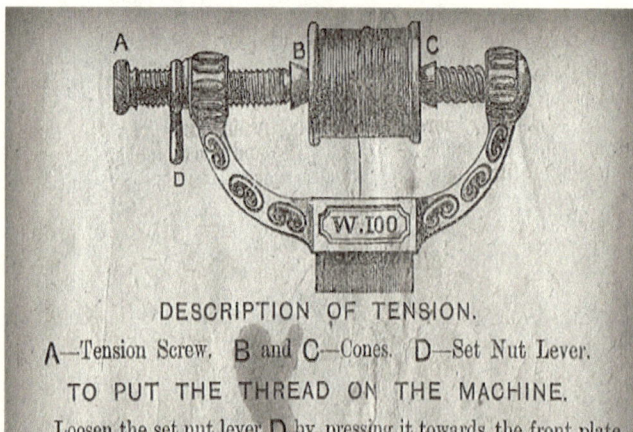

DESCRIPTION OF TENSION.

A—Tension Screw. B and C—Cones. D—Set Nut Lever.

TO PUT THE THREAD ON THE MACHINE.

Loosen the set nut lever D by pressing it towards the front plate and unscrew the tension screw A, insert the reel of cotton between the cones B and C so that the cotton rolls from the top of the reel towards the front plate and tighten screw A. When the required tension has been obtained tighten the set nut lever D. (By tension is meant the strain on the thread when coming off the reel. To put on more tension is to make the reel run harder by turning the screw A tighter against it; less tension, to make it go easier by turning it back.) The set nut lever D is to keep the screw from turning, and it is positively necessary to have it turned tightly against the spool holder.

Put the thread through the eye in the front plate and thread the needle from the front, then take that part of the cotton which is above the eye in front plate and wind it once round the cramp post B, bringing the thread where it is double on the cramp post, on the side next the cramping roll.

After close examination of many models over the years, let me tell you what he did improve upon over Raymond's model, (this bit is for the nerds among us, me included).

Weir improved not only the tension, which was now all incorporated, so that it did not fall apart every time you changed a reel of thread, but also the needle bar slide, which became wider.

He added a thumb nut for regulating the stitch length. The machine became far more practical and easier to adjust.

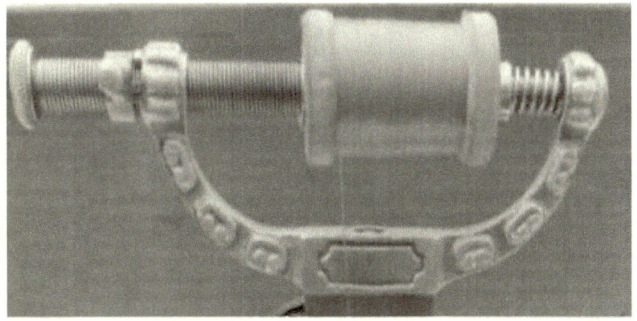

Weir's earlier patent (in 1872, Pat No 580) was for machine improvements. On February 23, 1872 Weir patented the spool holder. Further improvements were made in 1873 (No 2738). The improved patented thread tensioner on later Weir machines. Weir called it his patented tension, the Intermedial Tension. Note the space for a serial number, AT LAST!

He also added more oil holes for longer life. Another clever thing was that Weir offered these improvements to all the previous models as well, so you could return your old machine for a refit to bring it up-to-date.

Silent spiral-helical cross-cut gears on the later Weir, smoother, better wearing and quieter. The wider gears spread the load better and lasted longer without chipping. Weir had probably first seen these gears on the German machines flooding into Britain. They worked on his chain stitch and lock stitch machines perfectly.

James Weir Sewing Machines,
Easy Terms of Payment
Bankers, London & County Bank,
Oxford Street,
London

All in all, he did a great job on improving a bestselling machine. Then there were the boxes in different woods and some with little hidden drawers.

Weir's marketing skills kept his small chain stitch a bestseller, even though it did not do a lock stitch like many of the opposition's machines. It was the size, weight and price that made it so appealing and of course its simplicity. Even today there is no machine made as easy to thread as Weir's little marvel.

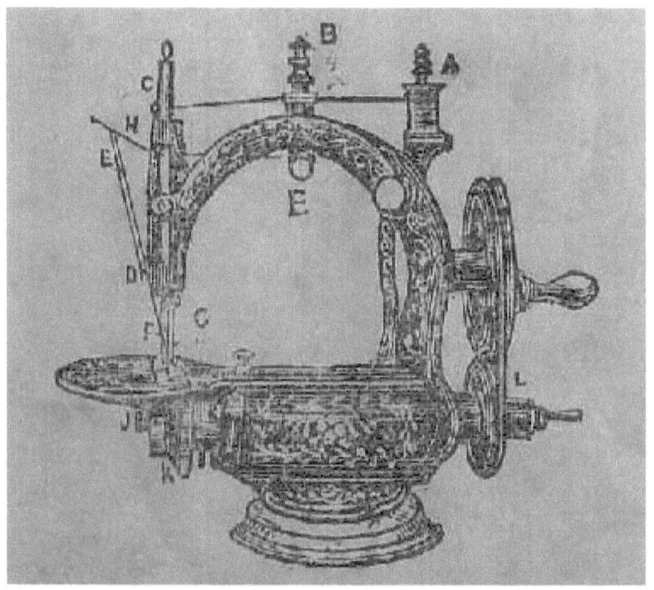

A woodcut of the super rare Weir Victoria, very similar to the Taylor Twisted Loop but supplied to weir by William Jackson of London.

All James Weir machines
Come with a 10-year guarantee!
Send to 2 Carlisle Street, London.
Machines will be returned by following post.

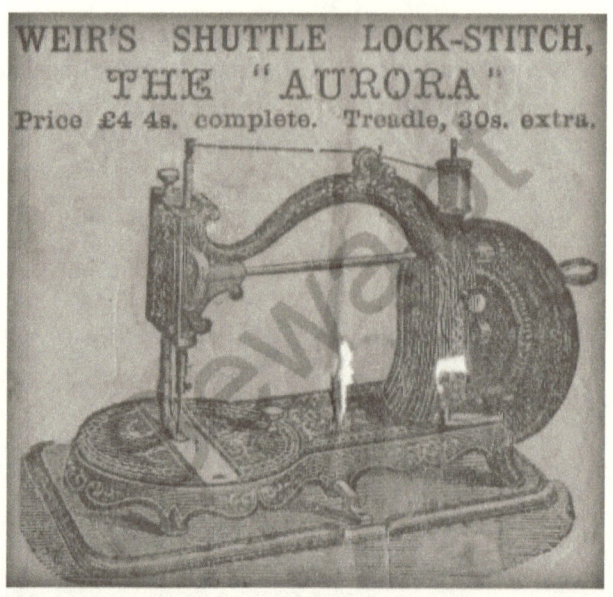

In 1990, I visited a woman who sold me a Weir sewing machine. Actually, I swapped a mornings work on all her sewing machines for the Weir.

Well, 24 years went by and in March of 2014, I called on her again. She reminded me that I had bought the Weir from her and then pulled out the original advertising leaflet that came with it. She has saved it for me for all those years.

Amazingly there were two sewing machines in it that I had not seen before and here they are, the Weir Aurora sewing machine and the Weir Comet sewing machine. The Aurora looks like a New Home family and the Comet either a Jones or Bradbury.

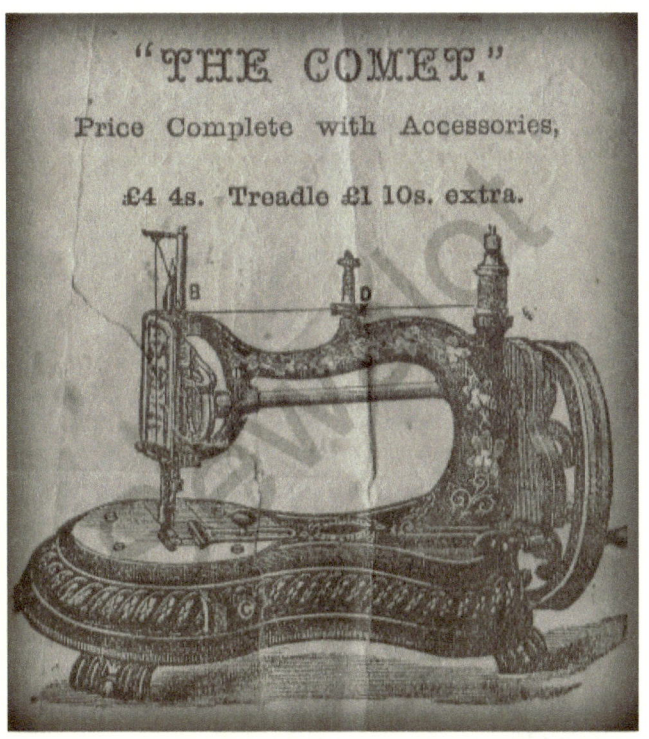

The Comet was a family lock stitch that retailed to the public for 84 shillings. By the 1870's there were so many chain stitch copies on the market that Weir became paranoid about making sure 'his' was the only machine to buy. This makes me laugh, as he was the person who originally copied Raymond's machine!

It went to the extreme when even his instruction leaflets became invalid unless they had been red-stamped genuine! All the literature that I have seen from the period of 1877 onward, clearly states that, unless the machine was bought from 'his only premises' at No 2 Carlisle Street, Soho, London West, they were not genuine!

He also mentions his address is two doors from Soho Square, just to make sure you don't buy a machine from one of his close competitors. That's a canny Scot for you!

The last Weir machine, a super-rare but plain Weir Argus Lockstitch. I have just one in my Sewalot Collection. £4,4s. If you wanted the treadle base it was 30s more.

Some experts say that the Weir Argus was a German import from Bottcher in Berlin. However, the similarity to the American New Home models of the same period is startling, especially New Home's Nelson model.

We know Weir was importing from America and to top it all, if I look closely on my model, in the right light, underneath the gold, you can just see the name Nelson across the machine! I bought my Argus from a dress shop where it was on display. It took a month of bargaining for them to let me have it but my persistence eventually paid off.

It is the only Weir Argus sewing machine to have surfaced around me so far! Although I know of a couple of Todd-Nelson's which are obviously from the same manufacturer.

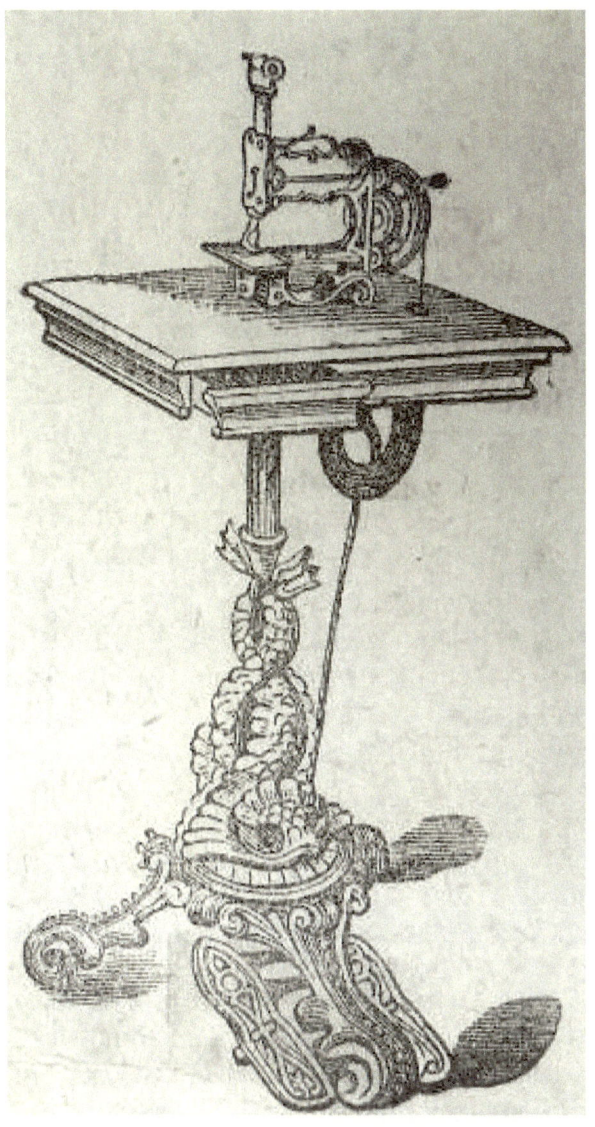

This is an actual Weir treadle, one of three models. It makes me dribble just looking at it! It cost an extra 30 shillings in 1870. The one I have in my collection took me 30 years to get.

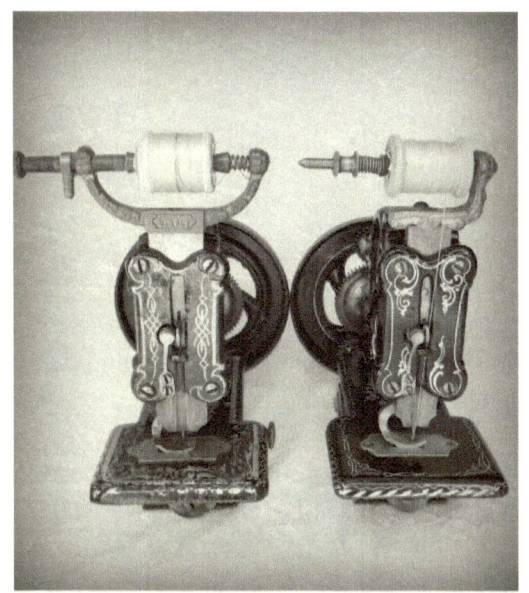

Here you can see the wider needlebar slide on the Weir and the improved thread holder.

Chapter Twelve

After Raymond and Weir had split, remember they had been mutually beneficial for years in Europe (and it was Weir who had established Raymond machines in the early 1860's), Raymond went full steam ahead sorting out new agents.

Raymond's agents included William Moore in Ireland and P Frank in Liverpool (who was also an agent for Richard Mott Wanzer). W B Moore had retail outlets in Dublin, Cork and Belfast.

Interestingly in 1870 Moore was selling Raymond's basic chain stitch and Weir's improved chain stitch at the same retail outlets! I have seen Raymond machines that had a W B Moore patent across the top spring plate. It's possible that he was also making improvements to the Raymond machines himself.

P. Frank was established in 1863 (the same year that Weir set up) and became Raymond's main importer after the acrimonious split with J G Weir. Frank sold Raymond's chain stitch at 55 shillings (the same as the better Weir improved chain stitch) and the Raymond New Family Lock stitch for £3. By selling Raymond's machine at the higher price he was increasing his profits.

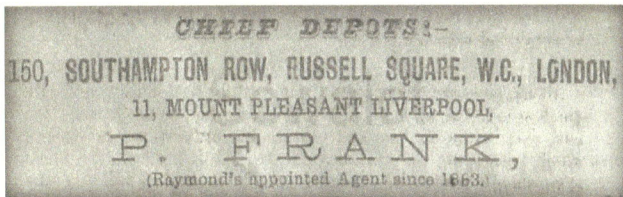

Frank, in Liverpool, was an importer for Raymond and other sewing machines from 1863. Frank stayed loyal to Raymond machines and constantly advertised against non-genuine machines!

Raymond now sold his London machines through the Highbury Sewing Machine Co of 75 or 73 Holloway Road North, London,

However, to make things worse for Raymond, by the late 1870's Raymond's production in Canada was in trouble, (as were several other Canadian sewing machine companies).

With the American Civil War over, competition from the huge American manufacturers on their doorstep was proving too much. A recession hit Canada in the late 1870's that lasted a decade. This was at the worst time for Raymond as he had now lost his main agent in Europe. Even worse, Weir was now in direct competition! It was a battle played out along the high streets of Europe.

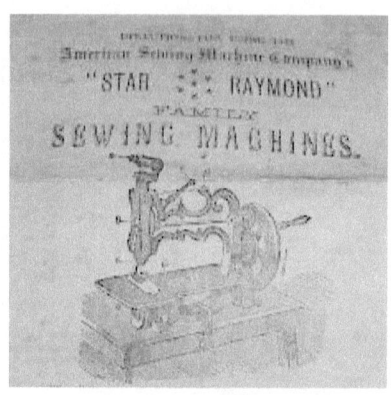

The Raymond Star sold in Britain during the 1870's. Probably to the disgust of J G Weir after their break. However Raymond's new lock stitch machines, with patented improved shuttles, were now being produced. He now had better machines to offer than just his old chain stitch.

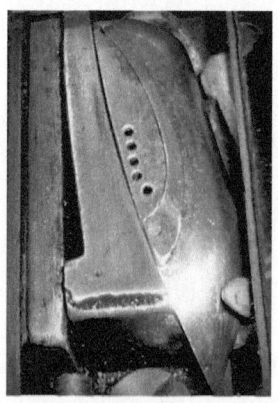

Weir had seen more and more agents popping up selling Raymond machines and this was cutting into his profits. Then Whight & Mann joined in.

*Whight & Mann
NOW agents for
the famous 'Raymond'
Little Darling Sewing Machines*

Charles Raymond secured the prestigious makers, Whight & Mann of Holborn, London, as his agents. Whight & Mann had sold the beautiful Prima Donna sewing machine. Here is one from my Sewalot Collection. They are so gorgeous they take your breath away.

He sold his standard Raymond chainstitch machine to Whight & Mann, infuriating Weir, who had premises just around the corner in London.

I have to wonder if James Weir didn't have a sneaky peek in Whight & Mann's shop window on the way to work!

Raymond was so angry with Weir that he even went so far as to take out adverts in trade magazines, letting everyone know that they had broken up, and where to secure his 'original' Raymond machines.

From this point on in the 1870's, Raymond and Weir were two separate sewing machines makers.

A few of my Raymond-Weir machines. They make an eye-catching display in my Sewalot Collection. The first model is an 1859 walking foot special.

Chapter Thirteen
The French Connection

The next two chapters have been the hardest to write. In fact after I had written them they were such a mess that I deleted the lot.

The problem is that there is no hard evidence where James Weir secured his Raymond copies. He was like the Scarlett Pimpernel, here there and everywhere. There are rumours of German, French, Irish and English makers, but not one single piece of paperwork to corroborate the evidence. Like William Shakespeare we know his work, we can see it everywhere BUT almost nothing remains of his original writings. They are just whispers on the wind. Weir is much like that.

We know business was booming for James. We know he had lists of eager customers waiting for stock. We know he was desperate to fill orders. What we don't know is where (besides Raymond) he got all his supplies from.

So I deleted the next two chapters. Then out of the blue I was sent two images, one of Seeling's of Paris and one of Vigneron. Suddenly, facts were emerging from the mist. Maybe I was on to something! So I decided to put them back just in case more evidence turned up to help future researchers. I wouldn't mind at all if you skipped the next two chapters!

Henry Vigneron, 50 Rue Du La Folie-Regnault, Paris.

After his split with the Raymond Company, for pretty obvious reasons, James Weir quickly found a French manufacturer who was already making bases for his machines. Apparently, Vigneron quoted Weir 26 shillings per machine, to supply the complete chain stitch.

Weir had possibly persuaded the French base makers to produce parts for his popular chain stitch model for some time. This backfired though, for a short time later Vigneron was selling a Raymond copy right on Weir's doorstep!

Like I say, out of all the machines on the market the Raymond was the easiest and simplest to reproduce.

Inadvertently, Weir had set up competition in Paris to his 55-shilling Dream Machine! It was not long before these copies started appearing on the streets of Britain. They are hard to distinguish from a genuine original Raymond or a later Weir copy of the Raymond. However, there is a little giveaway!

This brilliant photo is from the Rijnko Fekkes Collection in the Netherlands. Rijnko kindly sent me a picture of these super rare twins. You can see the Seeling sewing machine next to the Vigneron sewing machine.

Makers could produce a machine which was pretty much identical in every way. Even though Weir now had the ability to manufacture, it was easier and often cheaper for him to just buy in stock from other makers, to meet demand. This is something that he had been doing since the start of his sewing machine career.

La Favorite des Dames'
Machine a Coudre 'Raymond'
La Canadienne
H Vigneron
70 Boulevard Sebastopol
Paris

This is one of the French made Raymond copies by Henry Vigneron, very similar to Raymond's machine with the added bonus of an adjustable central gear to mesh the top and bottom gears perfectly.

I believe that only the French models (built by Vigneron) had this rear adjustment screw, so you can tell which ones are from France and which ones Weir made.

V H Seeling, Paris

A rare Seeling base to a Raymond machine kindly sent to me by Odile.

Another manufacturer that Weir used, before Vigneron was V H Seeling of Paris. He was also possibly in talks with Ms Goodwin of Paris, but I have never found evidence of a Goodwin chain stitch (yet).

Seeling's were making bases for Weir and later may have made the whole machine, but we don't know for sure. What we do know is that they sold **'The Ladies Favourite'** sewing machine. This could have been a Canadian import to start with, which they later copied!

The only firm connection I have so far was kindly sent to me by Raffaello in Italy, he has an early Chas Raymond sewing machine with a 'Seeling's of Paris' base in his collection.

Once again, we are putting together a jigsaw puzzle with many parts missing.

We know that when Seeling died his widow married Henry Vigneron. Henry took over Seeling's business and by the late 18870's, was producing his own sewing machine. He was also importing machines and adding his own base like the one above. Coincidentally this base would fit some of the early New Home machines as well.

'La Favorite des Dames'
Machine A Coudre Raymond
La Canadienne
H Vigneron
70 Boulevard Sebastopol
Paris

We know that after Henry Vigneron took over Seeling's business he won a big pay-out, from a law suit against Wheeler & Wilson and started seriously producing his own sewing machines.

By 1884 he was in full sewing machine production.

Compagnie Francaise des machines a coudre Vigneron, Paris

As I mentioned, W B Moore in Dublin were also making parts and bases for the James Weir machine. However, William Moore's bases, though super rare, are quite rough in their casting and it would be a big leap in quality to produce the Weir 'dream machine' fit for a queen!

I have been searching for definite proof of Weir's suppliers for 30 years and have yet to discover any rock-hard facts. I have some addresses that Weir used but he would keep his suppliers quiet in case someone did exactly what he did, cut out the middle man and get supplies direct!

Some of the big High Street retailers, do exactly the same today. They place huge orders with their suppliers. The suppliers turn over all their production to one customer. Suddenly that customer cancels all orders with the business. Of course, it goes belly up and who is there to buy it for tuppence! Now the High Street retailer has its own manufacturers. So simple and it works time and time again in big business.

It is not impossible to prove yet (but I'm pretty sure) that James Weir got one or more firms to copy the entire Raymond chain stitch.

Of course it must have been somewhere along the line that Raymond found out Weir was in the process of getting his own machine (in a country where Raymond did not have patent protection) and that kicked the whole ruckus off! Certainly by 1873 Charles Raymond was aware of Weir's modified chain stitch.

For now, Henry Vigneron is the most likely French manufacturer to have been making and supplying Weir. Only time will definitely tell.

Seeling's of Paris, later H Vigneron, made this super rare cast base and possibly supplied entire machines for James Weir at a later date.

With money from his lawsuit against W&W, Henry Vigneron started producing large quantities of his own machines.

By 1884 he was producing nearly 8,000 machines a year. Check out the next advert for Vigneron. You can clearly see that it is a copy of a Wheeler & Wilson No9.

The problem for American companies were the French took it for granted that they had in fact invented the sewing machine with Barthélemy Thimonnier in the 1830's, so any machines were copied from the French in the first place!

After Vigneron's time, A. Total took over selling two models in the 1890's, the Raymond/Weir chain stitch copy and the W & W lockstitch copy.

Machines ready for shipment to the colonies complete with written guarantees at one hour's notice!

The Vigneron copy of the Wheeler & Wilson No9. Agents, A. Papin, Rue des Halles, Tours, France. Vigneron and his successor, A. Total, successfully sold this model for several years. Their win over the giant Wheeler & Wilson Company was a real poke in the eye.

Chapter Fourteen

Wow, I'm glad those two chapters are behind us.

Time rolls on and Weir is now getting tired of the business. He has spent 20 years playing with sewing machines, including a superb machine he called The Victoria (possibly an early Wanzer) but none of these sold in the quantities of his chain stitch.

Weir was juggling so many companies and so many importers, it must have been mind-blowing trying to keep up. Of course, he was still only 40 so he had the energy to manage it all. To make matters worse, several copies of his chain stitch were appearing on the market. The Express was selling for less than half the price of his machine! The New England chain stitch had now become one of the most copied machines on the market!

The Automatic Machinery Company

Weir also had a storage/manufacturing facility he referred to as '*his works*' at Ferdinand Place in the Chalk Farm area of London. Interestingly, at Ferdinand Place, Weir was also involved in a business called The Automatic Machinery Company. It is possible that it was here that some of his patents were put to use.

In 1878 the company folded. It was not long after this point, and the previous loss of another business interest, The British Boot & Shoe Machinery Co, that Weir took the decision to go into politics full time.

This was a stroke of genius. To walk away while he could. So many other makers reinvested all their hard-earned wealth and lost the lot. At Weir's peak it was publicised that he was selling over 18,000 machines a year. Now with competition on every street corner and sales collapsing, it was time to pack up and head for the hills.

> 2, Carlisle Street,
> Two doors from Soho Square (West side)
> London, W.
>
> Madam,
>
> I beg to inform you that the 55s. American Hand Sewing Machine is now, and in future will be kept in Stock, and can be supplied at a moment's notice.
>
> Soliciting a continuance of your recommendation,
>
> I am, Madam,
> Your obedient Servant,
> JAMES WEIR.
>
> N.B. A liberal reduction on 3 Machines and upwards, to Charitable Institutions.

A very rare note from Weir advertising his New American. Now in my Sewalot Collection.

Chapter Fifteen
A life in Politics

At 41, James was fed up with the cut throat sewing machine business, and the constant competition (new machines were turning up by the day). His patent protection was also running out. Soon anyone could copy all the early ideas. His machines were under threat. The glory years of his dream machine were behind him. Time to get out and follow his love of politics!

We all know how that feels, you know, that the grass is always greener! To cure that problem, buy your neighbours garden then the grass is yours on both sides!

James Weir decided to retire from the sewing industry around 1879-80 and follow his passion. How we would all love to retire early. I think I will drop dead over a sewing machine at 75 with some old dear prodding me with a stick to finish the job!

Anyway, James Weir left sewing machines behind him, passing it on to his old colleague J W Columbine. Time for politics!

In 1892, James Weir, after an earlier failure in Falkirk in 1885, was elected a Liberal Member of Parliament, for Ross and Cromarty. He followed a colourful life in politics for many years.

Lots can be found about his career, but we are concerned with his sewing machine life so I will desist from too much waffle on the subject.

Although they say he was not the strongest politician, he was full of energy and determination. He doggedly supported the rights of crofters in his constituency, and worked tirelessly in Parliament, and was said to be one of the most active members in the House of Commons.

Now, I have a little further information to add about James, from Dawn Siggs. Dawn was born in Brisbane Australia, in 1936, and sought me out to tell me of her distant relation. After a lovely chat over a cup of tea she added some great information. Remember James Weir's first wife, Mary Anne Dash, from Brighton, Sussex?

Well Mary Ann Dash was a furrier in Brighton and met James Weir while he was a travelling salesman for a haberdashery firm. Before he started importing and selling sewing machines he called regularly on her during his supply rounds.

This is a really interesting point for it shows how James became involved with sewing machines. He was supplying the very trade where he knew his market was.

They had a son and three daughters before her death. Little is known about his son. The daughters were Edith, Alice and Amy.
Amy, an apothecary, married a doctor from Stornaway and moved up to the island. She once organised a group of Stornaway crofter's daughters

to travel down from Scotland with her father, and to sing local crofters' songs to the House of Commons. They sung to the entire Commons with much applause. James was always a staunch supporter of the poorer tenant farmers.

Both the other sisters moved abroad, one to Paris where she may be buried, and the other to Italy. She possibly died after an earthquake. Apparently the Winterborn Family still have her diary. Amy died in 1910, three years after her husband, they had no offspring to my knowledge. It is possible that Amy died of TB and is buried in her father's plot in Marylebone.

I may be wrong on some of the details and would love to find any distant relation to James to correct these snippets from Dawn.

Weir machines, copied from his Canadian counterpart Raymond, were just beautiful. You can see why they had so much appeal.

Now I need to be corrected on this leap...Much later in life as an old man, James Weir married again to Marion Jolly, from Northumberland who is buried in Milverton, near Kenilworth. They had two children, a girl Margaret, and a boy James.

Now, I'll tell you something amazing. Many moons ago I bumped into the wife of Andrew McLaren Winterborn (J G Weir's grandson).

I had the pleasure to visit Frances Winterborn who had called me out to service her sewing machine in St Leonards, East Sussex. In her living room was an oil painting of a grand old man that looked so familiar. I kept staring at it but could not fathom why it felt like I should know him.

When I asked who it was, I was amazed to be confronted with one of sewing histories giants, none other than James Weir himself. I promptly got my camera from the car to take a picture. That was a surreal moment.

It was always assumed that when James retired to follow his yearning for politics, he wound up his sewing machine business. It was only many years later that I came across a receipt from Columbine in the 1890's, showing machines with the Weir name were still being sold years after Weir had retired.

After a bit more digging, I found out that James Weir had handed over the firm to James William Columbine. Columbine and Weir had many business dealings together over the years, so it is possible he kept some quiet financial interest in the business after his retirement. William Wood then continued with Weir machines in the mid 1890's.

> **JAMES W. COLUMBINE**
> *(Late JAMES G. WEIR),*
> SEWING MACHINE MANUFACTURER
> TO HER MAJESTY THE QUEEN.
> WEIR'S 55/- SEWING MACHINES.

This little piece of paper is the only one that has surfaced so far. It is part of an original receipt in my Sewalot Collection for a machine for Mrs Cooper, sold to her by A E Columbine. A E Columbine took over after James W Columbine, who took over from James Weir. The receipt is dated from the 1890's and has their new address after moving from Weir's Carlisle Street, Soho address.

It shows two very important things. Firstly they were still proudly boasting patronage from Queen Victoria. And secondly in the 30 years or so since James Weir had started selling his chain stitch machine, the price had stayed the same. Amazing!

And here is something special, a super rare Weir Improved Argus in stunning condition. Weir imported many of his machines. The most likely maker of this beauty is the German company Bottcher of Berlin.

Before his retirement Weir had experimented with a few other machines. The last machine with his name stamped on was the Improved Argus lock stitch. It sold for the sum of 84 shillings and was their most expensive machine. Columbine, and later Wood, traded from the same Soho shop, selling an assortment of sewing machines, until that finally closed in the late 1890's.

James G Weir died at home at Frognal in Hampstead, after suffering a stroke in late spring, 18 May 1911. He was in his early 70's. He is buried in Marylebone Cemetery, London.

His wealth by this time was said to be considerable. His reputation high. He was held in great esteem by his friends and political adversaries alike.

One of the giants of the early sewing machine industry had gone but what a legacy he had left behind. Some of the most sought after and collectible machines of all time. Every serious collector should have at least one in their collection.

I have never tracked down who did this wonderful image. It was sent to me many moons ago and has been on my Sewalot Site ever since.

Chapter Sixteen
Back to Charles Raymond

Our extraordinary story of James Galloway Weir is now done and dusted so we can now return to Charles Raymond and finish his journey.

We need to back-peddle a little to the year 1871. By 1871 Charles Raymond had been a busy boy. A big factory with a mansion opposite to walk across the road from each morning. Lots of new ideas and new machines coming on line, and at last some proper lockstitch machines, models 1, 2 and 3. All basically the same machine but increasing in size for light to heavy work.

Interestingly Charles Raymond never copied his main Canadian competitor's machines, Richard Mott Wanzer. This showed the character of the man.

Lockstitch shuttle Patent No 1433

In April of 1872, Charles Raymond applied for his Improved Lockstitch Sewing Machine Patent. Later that year he was granted his first Canadian Patent for a lockstitch shuttle machine, No 1433.

Interestingly in 1873, the Raymond Sewing Machine Company Employees Mutual Benefit Society was set up, to look after the workers at the Guelph factories.

This was great for the workers, especially if they had an accident. It was benefits like this that kept the workforce happier and production targets met. Charles understood that it was not only the wages but his workers health and wellbeing that was important.

A rare Raymond trade card for the New Raymond circa 1878, (courtesy Mike Smith). Towards the end of Raymond sewing machines many names started appearing on the arm of their sewing machines, it is possible, like the Jones Company in Britain, The Raymond Company would put the name you wanted along the arm, if you bought a certain number of machines. The Economy Model, produced from the 1890's, was the perfect example of this.

Back in 1875 disaster had struck with a fire destroying most of the old wooden works. A spark from the foundry caught in some old straw that had been badly cleared from the stables (remember there were lots of children who worked at the plant). Wind blew it onto the old dry wooden buildings. Before long the entire works were consumed in flame. By the time his works manager, John Sully, came flying into the restaurant, where Charles was eating with some of his fellow Guelph politicians, the factory was little more than rubble and ashes. I guess that gave him heartburn!

Attachments for Raymond Sewing Machines

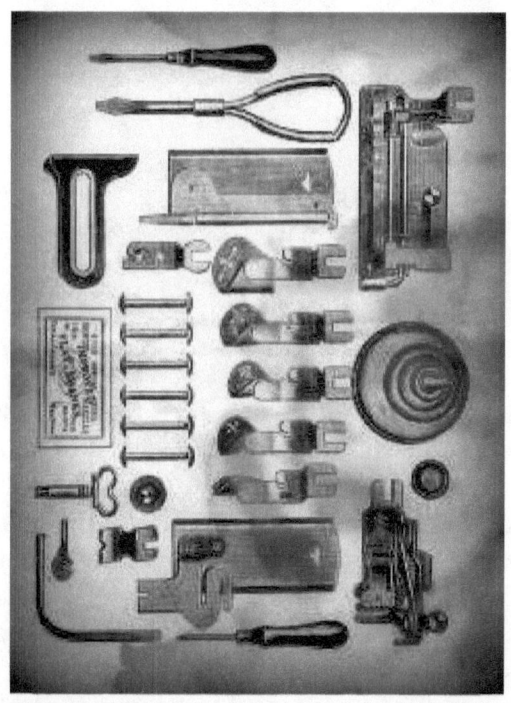

Charles Raymond was not a man to be knocked down easily. He acted quickly, buying out another local sewing machine competitor that had started up, the Arms and Worswick Co. He used their premises and staff while rebuilding his old works, this time with fireproof brick! Amazingly by the end of the year production had only fallen by 27%. Within two years it had not only matched previous years, but stormed by it.

As production from his new factory and foundry came on line, his latest machine was a beauty.

The Charles Raymond High-Arm sewing machine first came onto the market around 1885. Notice the CR (Charles Raymond) logo on the plate next to the Raymond name, this changed to RM (Raymond Manufacturing) in 1895, just before he retired in 1897. If your machine is marked CR it is probably pre 1895 and RM, Raymond Manufacturing, post 1895.

The Raymond Manufacturing Company Drop Head and ornately carved treadle.

Chapter Seventeen
Retirement Looms

All this pressure took its toll and in 1877 Charles Raymond suffered badly with his health. He decided to have a break from factory work.

The factories were now employing over 200 men 11 women and 28 young lads. The business was well run and the factory managers almost ran the place when Charles was hit by his fatigues and illnesses. Charles had no worries when he was away from work. The business was in the excellent hands of John G Sully and Christian Kloepfer.

John Gilbert Sully knew the business from the ground up. He had started as a basic labourer and slowly worked his way up through various departments in the factory. Finally taking hold of the company reigns, as General Manager. He knew every job in the factory backwards and inside out. If a worker came to him with a problem, he understood and knew the solution. Sully died in Toronto at the ripe old age of 87.

22nd Sept 1878
Winder improvement patent granted

Spells away from his factories allowed Charles Raymond time to play and, as a keen politician, inventor and philanthropist his mind was always occupied with new ideas. He could also seek out new agencies for his machines around the world.

> *J F Sherlock, General Agent for*
> *Raymond Sewing Machines*
> *222 Dundas Street*
> *London*

In 1879 Charles Raymond and a couple of his friends arranged the installation of Guelph's first telephone exchange, remarkably early for the time. Once again this was a great idea and very useful for Raymond's business. Mind you in 1879, who were you going to talk to?

Charles Raymond was amongst the leading 'Guelph men' of the day, he had a hand in most important decisions that influenced the growing town. For example, he was a member of the school board, building the Central School, a member of the hospital governors (where he later died) and even the County Poor House.

He was a main investor in the Guelph Iron Horse (as a director of the Guelph Railway). Of course, this was instrumental for his business. An efficient railway was essential for bringing in raw materials and sending out sewing machines. The railway station was the first and most important stop on the journey that his sewing machines would take. For some, it was the starting point that would lead to the four corners of the Earth. By now Raymond machines were known from Alaska to Australia.

In September of 1879, the fourth daughter of Queen Victoria, Princess Louise, and her husband John Campbell, the Marquise of Lorne (Governor General of Canada) stayed with Charles at his home. They had come to open the Central Exhibition in Guelph. From then on Charles' impressive home became known as Lornewood.

In February of that year Princess Louise and her husband, the Governor General of Canada, had opened the Canadian Parliament.

To crown Raymond's best year of 1879, Charles perfected his finest lock stitch yet, the fabulous Raymond No1. The No1 was a beautiful looking machine and ran for around six years.

It is interesting to note that by 1880 Charles Raymond was still only making a few distinct sewing machine models but they were now making around 600 complete sewing machines a week!

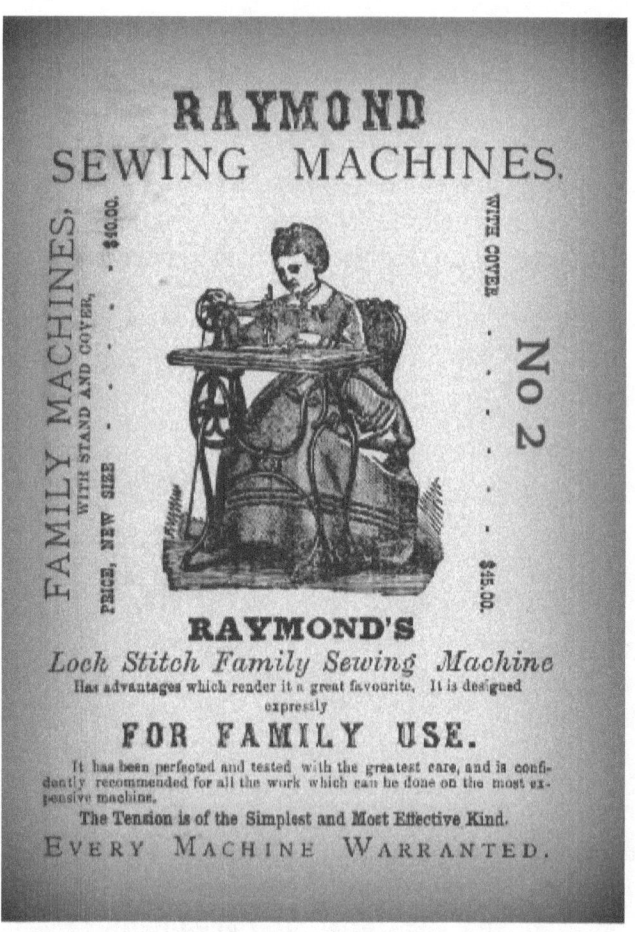

The Raymond Family No2, soon came along, a bigger better machine. This model was suitable for professional workers as well as the home. Retailing at $45 complete, or cheaper with no lid, you can see that the company was always trying to cut costs. They had some big competition in Canada and across the border in America.

7ᵗʰ May 1880
Patent approved for
Further improvements
To the bobbin winder mechanism,
Raymond Lock Stitch.

In 1886 Raymond became an alderman. He belonged to the Guelph board of trade and Canadian Manufacturers Association. He oversaw the construction of several important buildings in Guelph, many still standing today. Deeply religious he was a deacon of the Guelph First Baptist Church and read sermons regularly there. It was with his help and donations that the church on Woolwich Street was built.

Charles Raymond was a friend of the community in the town, and an active member of the Guelph Board for Foreign Missions. All-in-all an opinionated but thoroughly decent human being. Someone the city could be proud off.

"It is to the personal efforts of
Charles Raymond
that the people of Guelph are largely indebted."
The Guelph Herald 1880

It is interesting to note that there was almost no notable landmark in Guelph that Charles did not have a hand in. No wonder he was worn out!

However, he hated drink and no one who worked for him was allowed to drink! He even abstained from a celebration tipple. In fact, he once ran for local office on a platform of alcohol abstinence. He lost badly. Obviously, the other Guelph inhabitants did not quite see eye to eye with him on his moral crusade.

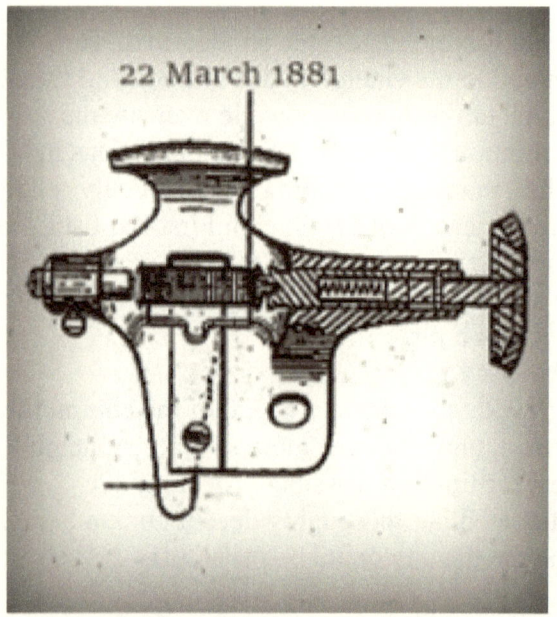

**Patent 239059 March 1881
'Sprung' bobbin winder, Charles Raymond.**

The Raymond Manufacturing Company Limited

In 1895 Charles Raymond incorporated his business to become the Raymond Manufacturing Company Limited. This allowed other goods to be made.

The Charles Raymond High-Arm sewing machine. Notice the CR logo on the plate left of the Raymond name, this changed to RM in 1895, just before Charles Raymond retired in 1897. Remember a machine marked CR is probably pre 1895, RM post 1895.

Raymond Manufacturing Post 1895

By 1897, Charles Raymond had decided enough was enough. He was getting old and in failing health so at the age of 71 he stepped down from the sewing machine business, passing it on to his faithful factory manager Sully, and his old partner Chris Kloepfer.

From this point on in history the company was named The Raymond Manufacturing Company. Why? Because Sully and Kloepfer had another project in mind, as far away from sewing machines as you could possibly imagine!

You would think that Charles Raymond would put his feet up and take it easy. I wonder if, at his mansion opposite the factory, he used to sit in the garden watching his old business, while puffing on his pipe.

I know that although he retired, he still kept playing with his sewing machines and other ideas, including the latest new craze of petroleum engines, right up until his death in 1904.

Charles Raymond's monument is within the Woodlawn Memorial Park, Woolwich Street, Guelph.

Charles Raymond was 78 when he died (in the same month of his birth) following a failed operation on his bladder.

It is quite possible that Charles knew his time was near for he gave most of his wealth away in several bequests before his death. He died with almost no riches, just a few thousand dollars and his grand home.

Despite new models, without his inspiration and guidance, the Raymond Manufacturing Company floundered.

Interestingly, the company was temporarily saved by diversifying into cream separators.

Chapter Eighteen
National Cream Separators

The Raymond National Cream Separators were the best on the market. The skills used to create complex sewing machinery were put to good use and the separators flourished. The hand models could separate over 30 gallons of cream an hour. The powered ones, double that!

> *The Perfect Skimmer*
> *The world renowned*
> *National Cream Separator*
> *By the*
> *Raymond MFG. CO Limited*
> *Guelph Ontario*

It was boasted by the company that by 1906 eight out of ten cream separators in Canada were made by the Raymond Manufacturing Company. Agents were hired and supplies were sold throughout the country.

All this was to the expense of sewing machine production, which had been falling steadily for years. In 1891 only 3,000 machines had been made! Numbers had increased for a short time under White Sewing Machines but, by 1922 only six sewing machines a day were being made!

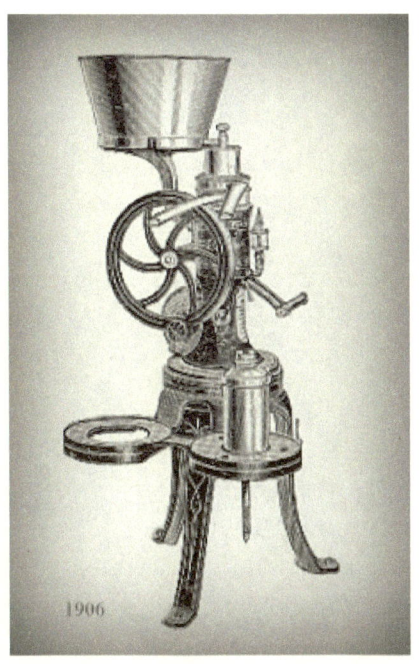

The National Cream Separator by the Raymond Manufacturing Company, saved the company for several years as competition from around the world ate up their sewing machine market.

However, the company was now big and heavy with a lumbering workforce. It needed huge expense just to stay open, let alone make any profit. The other problem was that, since Charles had retired, they had not kept up with modern technology.

Later model Raymond sewing machines were all pretty similar with slight variations to the decoration and winder assembly. The quality was great but that came at an ever-increasing cost.

After Charles Raymond retired, his old partners ruled the roost. Their cream separators took more and more of the production. This image, possibly from the late 1890's, shows the main factory in all its advertised glory. The darker building behind the main one was now exclusively for the manufacture of the popular National Cream Separator. The problem was once the machine was purchased (by a much smaller market than sewing machines) the demand dropped. Sewing machine sales were booming with giants like Singer producing upwards of a million machines a year. This switch of direction spelled the end for the Raymond Company. It became easy pickings for the big boys.

Chapter Nineteen
White Sewing Machine Company of Canada

This beautiful model Raymond sewing machine is with kind permission from Vern Schafer. It is in simply stunning condition, possibly one of the finest museum quality pieces ever found.

The White Sewing Machine Company
Of
Canada

In April of 1916, the Raymond Sewing Machine Co was sold to the giant White Sewing Machine Company that survives to this day.

They changed the Raymond Manufacturing Company, to the White Sewing Machine Company of Canada. I'm not sure what happened to their popular cream separators, but it makes sense that it was sold to another business.

The White Company immediately tried to increase production while cutting costs. They expanded the factory sites and put up prices of the Raymond machines. This led to a massive drop in orders. What a surprise!

They did little to improve or change the machine but did decorate them in a variety of different decals. Some stunning I might add.

For a period of six years, from late 1916 until 1922, Raymond Sewing Machines were marked as White Sewing Machines. Now if they were averaging around 600 machines per week, back in the boom years. We can surmise, that in the life of the Raymond Sewing Machine Company, from the middle 1850's until 1922, around 400,000 Raymond sewing machines and Raymond-White sewing machines may have been produced. This has led to a fertile collector's market for Raymond-White models.

In 1922 the White Company decided to shut down their plant in Guelph. That must have been a tragedy for the city coming so close to the end of

World War One. Today, some of the factory buildings still survive as apartments.

In all the years of making sewing machines, the original Raymond business produced only a handful of different models (with many different names). These machines went through minor modifications and improvements along the way.

For example, the first and most popular machine, the Raymond chain stitch, which went from the 1850's until around 1886, had no serial numbers to help with dating. It also had at least 11 different names, from the household Fairy to The American, then the New American. The Raymond Household made between 1869 and 1886, didn't have serial numbers either. Funnily, it is easier to date the British Weir models as Weir made regular changes to Raymond's machines as he improved the mechanisms.

This beautiful machine came from Mike at Wolfegang's just before I went to print. At first glance it looks like the No1 however it pre-dates

the No1 and is several inches smaller. Is this Raymond's lockstitch prototype that he first advertised in 1869? It's certainly a beauty and in stunning condition. Note it is a transverse shuttle based on the ever-popular Singer 12.

Here you can see a superb Raymond No1 from the Carl Koenig Collection. If you look closely at the bed you will see it is far wider than the earlier image.

Many of the larger Raymond models such as the No1, No2 & No3 did have serial numbers but no one has found specific records to help with dating. So, pinning down the years between when they were made in the latter half of the 19th Century is difficult. It can be done, but it needs a concerted effort by someone to start a survey and compare all the numbers collected from surviving machines.

When the Raymond Economy, High Arm and other models came on line from the middle 1880's, the names and models seemed to change with the weather, The Raymond, Chas Raymond, New Raymond, Victor, Canadian, White Canadian, Beaver, Windsor, Eaton and more. Then there were all the model letters as well!

From the 1890s until the closure of the factories in 1922, there were so many model changes (all similar machines) that it is very hard to follow.

However, don't give up hope. Enthusiasts are working as I type to put all the information together and create some clarity about dating later models. To date no Raymond has ever turned up with a serial number above 174000.

Chapter Twenty
Values

There are ways to value your antique sewing machine reasonably accurately, especially for insurance.

The value of anything on this planet always depends on what someone is willing to pay you. A splash of paint is just a stain. If it was done by Jackson Pollock, it could be worth millions. With sewing machines it is all about rarity and condition.

Rare machines, even in poor condition, fetch good money. Common machines in great condition also fetch good money. So how do you value the ones you have?

We are lucky to be in the age of the Internet. Before then I wandered around old junk shops and perused antique publications, hoping to fall across some ancient collectible. I would travel the width and breadth of the country going to auctions and antique fares. In the dead of winter I would be walking up and down London markets hoping to come across a single machine. When I did it was quite often rusty rubbish.

I would also advertise for early machines. That used to bring a few laughs, "This machine was used by Admiral Nelson." That sort of tosh! And then glory be, the Internet came along. Now we can search the world in a heartbeat.

If you want to find a true value of your machine you need to wait till one, in the same condition turns up on a popular Internet auction site. One that gets lots of visitors. Forget about 'Buy It Now' that's just the 'enthusiastic' being hopeful. What you need to do is see the **final auction price**. What the machine actually sold for.

The price can still vary. Bad images always get less. Good deals are always had from people too lazy to pack and post machines. They will always go for a lot less than the price they should have done. People who finish an auction at a silly time, or on a public holiday will also lose out.

What you are looking for is a machine like yours, finishing at a good time for all. That final price will be close to what the value of your machine is. Now, watch a few machines like yours finish, and average the prices. That will give you the current value of your machine today.

Raymond, Weir and New England style chain stitch machines will always go for good money. They are a collectors dream and appeal to not only sewing machine enthusiasts but toy collectors, even engineers.

A near perfect early model, with great images, on a good auction site, well worded and advertised, will nearly always fetch between $500 and $1,000. On a lucky day, that can double. The great thing with auctions is, if two, just two people really want an item, the sky is the limit. Of course you could always splash paint over it and call yourself Jackson Codfish or something.

Buying an old machine

Make sure if you are bidding on one, it has no broken or missing parts. Don't be afraid to contact the seller, and check their feedback score. You will find it hard to get any original parts for early machines. Also, in the 19th Century, many sewing machines were what was then called file-to-fit. Each piece unique and made for that model only. A similar part from an identical machine may never work!

Actually that's a final point worth mentioning. I adore the original burnished steel worn look of old sewing machines, as do many others. Wolfegang's Collectibles is the ideal example of that. Mike does not mess with his machines, they are shown in perfect detail in all their worn glory, and always fetch great prices. A badly touched up machine will drop in value like a stone.

Well that's all folks! Two more of the Sewing Machine Pioneers, done and dusted. I hope you enjoyed the fascinating journey and learnt a few pointers along the way.

All I need to do is live a few more years and I'll have the rest of the Sewing Machine Pioneers down in print forever. Bye for now.

The End

Chas Raymond
&
James Weir
Sewing Machines
By
Alex Askaroff

Sewing Machine Pioneer Series

To see other publications by Alex Askaroff, visit Amazon

Isaac Singer
The First capitalist
No1 New release

Most of us know the name Singer but few are aware of his amazing life story, his rags to riches journey from a little runaway to one of the richest men of his age. The story of Isaac Merritt Singer will blow your mind, his wives and lovers his castles and palaces, all built on the back of one of the greatest inventions of the 19th century. For the first time the most complete story of a forgotten giant is brought to you by Alex Askaroff.

No1 New Release. No1 Bestseller Amazon certified.

*If this isn't the perfect book it's close to it!
I'm on my third run through already.
Love it, love it, love it.
F. Watson USA*

Elias Howe
The Man Who Changed The World
No1 New Release Amazon Oct 2019.

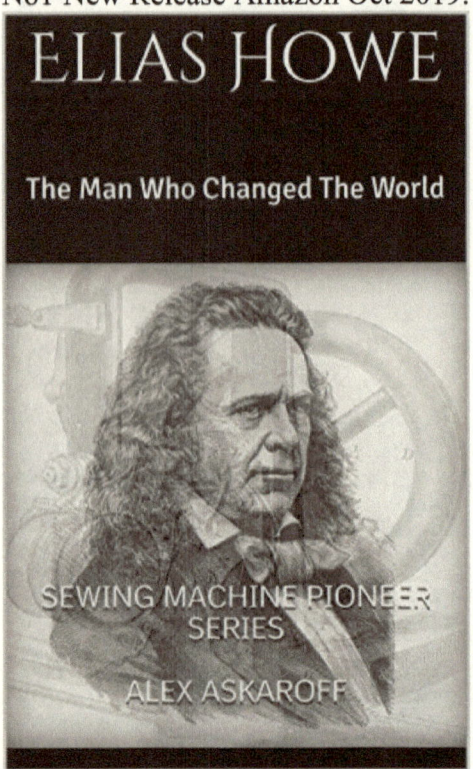

Anyone who uses a sewing machine today has one person to thank, Elias Howe. He was the young farmer with a weak body who figured it out. Elias's life was short and hard, from the largest court cases in legal history to his adventures in the American Civil War. He carved out a name that will live forever. Elias was 48 when he died. In that short time he really was the man who changed the world.

www.ingramcontent.com/pod-product-compliance
Lightning Source LLC
Chambersburg PA
CBHW020431220526
45464CB00002B/651